Machines That Think and Move: A Guide to the Future of Robotics

Jainson

Contents

1 Introduction

1.1 Motivation

„There certainly will be job disruption. Because what is going to happen is robots will be able to do everything better than us. I mean all of us."

Elon Musk

This statement about robotic application potential was given by Elon Musk during a press conference [18]. In general, a robot is defined as a machine, which is able to sense, think and act. Basically, robotics is a confluent technical development of sensors, advanced algorithms and actuators. Indeed, the sensors are needed to obtain information from environments. The advanced algorithms are adopted to make decisions autonomously, while the actuators are employed to exert forces resulting in robot motions in the environments [4].

In general, robotics has wide range applications, originated from classical manufacturing and industrial assembly facilities to other new domains [40]. For instance, a famous robotic Da Vinci system has been applied for medical surgery, which is able to allow two surgeons to operate together [89]. An ambitious DARPA program is another example, which aims to develop and revolutionize disaster response robots [35]. Overall, there are numerous unprecedented application potential such as in space exploration, automated driving and material handling [40, 49, 100]. In 2015, a global spending on robotics came about seventy one billion US dollars as shown in a statistics [106]. Due to their capability of replacing human labor and improving efficiency, robots have a huge market, which is expected to grow quickly in the next ten years [33]. However, there are great challenges in algorithms development. In fact, not only the number of robots grows but also more complex tasks have to be performed. Consequently, more advanced and efficient algorithms are required in order to solve these problems. Particularly, motion planning algorithm is the most important key, which enables a greater autonomy for robotic systems.

Motion planning is sometimes referred as trajectory generation, which is theoretically an extreme challenging problem. In trajectory generation, several factors must be taken into account. First, collisions with static obstacles as well as among robots must be avoided in order to guarantee safety. Moreover, trajectory performances are optimized for the purpose of utilizing robot capability and maximizing productivity. If the number of robots is large, cooperative trajectory planning is also a prerequisite. Therefore, coordination algorithms become essential. Motivated by these requirements and demands, this book is devoted to solve some optimal and distributed motion planning problems. In this book, several novel results and algorithms are presented.

1.2 Problem Description

As parts of production lines, material handling systems play an important role in industry. Conventionally, goods and wares are carried by one-dimensional linear conveyors. Possible paths are predefined for each item by installed equipments [87]. In this case, a continuous material transportation flow is generated. Moreover, collisions and deadlocks [1] are inherently prevented due to the continuous transportation flow. Nevertheless, technological and robotic advancements enable another type of material handling, which is discontinuous and more flexible. Indeed, each item can be individually transported in a two-dimensional space by a robot [29]. However, these discontinuous material handling systems induce a great demand for advanced algorithms, which are able to prevent collisions as well as resolve conflicts among robots. Consequently, several open research questions have been raised. Before making an excursus about these research questions, an overview about the discontinuous handling systems is provided and their characteristics will be specifically described.

Figure 1.1: Grid-flow with transport vehicles. Photograph of Amazon Kiva Systems [21]

Due to an industrial revolution 4.0 initiative, several modular flexible material handling systems have been recently developed and demonstrated in industrial shows. For interested readers, an exhaustive survey is conducted by Zäzilia Seibold [87]. Despite being known under different names such as plug and play, cellular and fluid logistics, etc., these systems can be categorized into two groups. The first group is a grid-flow with active transporting robots as shown in Figure 1.1. Such systems can be found in many warehouses, where pallets are placed in a grid pattern. Furthermore,

---

[1] *A set of processes is deadlock when every process in the set is waiting for a resource that must be released by another process in the set* [96].

the robots are able to translate underneath the pallets. Consequently, they are able to pick and deliver the pallets to any destination [21]. In such systems, each robot can decide its own movements. Moreover, conflicts are resolved by global central algorithms running on a master computer.

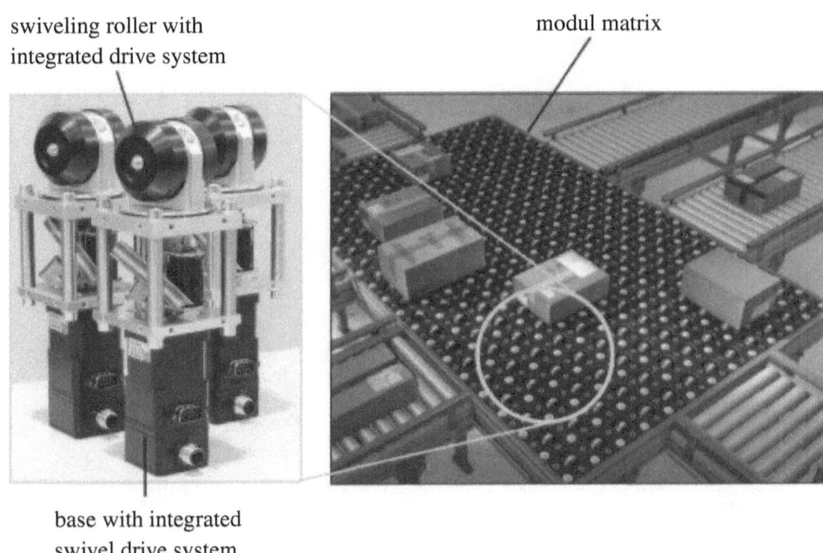

Figure 1.2: Photograph of a cognitive convey [50]. The omhniwheels are placed underneath items.

In contrast to the first group, robots are passive and driven by actuators placed underneath in the second group. The actuators are controlled in a decentralized manner. An example of such systems is depicted in Figure 1.2. In this example, the actuators are omniwheels. As opposed to the previous case, deadlocks and conflicts can be resolved by distributed algorithms deployed directly into the actuators. Nevertheless, very few results about the distributed algorithms are published, which are also not precisely described [87]. Moreover, these distributed algorithms can only be applied for primitive motions, e.g. four cardinal directional movements. In order to fully utilize the system flexibility and maximize the whole system performance, more complex and aggressive motions are desired, which results in the following open research problems.

Problem 1: Time-optimal Trajectory Generation Problem
Time-optimal trajectory generation is a fundamental optimal control problem, which is encountered in robotics. Given a list of obstacles, initial and goal states, computing time-optimal trajectories is a very complex problem. In fact, collision avoidance induces non-convex constraints.

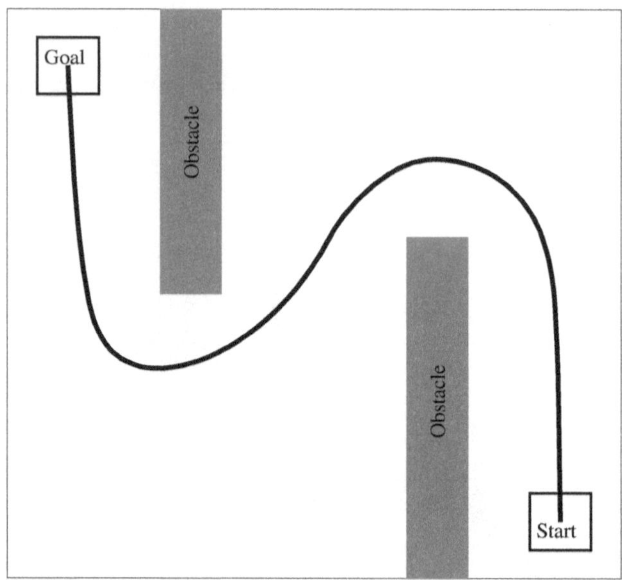

Figure 1.3: The time-optimal trajectory generation problem

Basically, the time-optimal control problem is formulated as follows

$$\min_{\mathbf{q}(t),T} \quad T \tag{1.1a}$$

$$\text{subject to} \quad \mathbf{q}(t) = \begin{bmatrix} q_x(t) \\ q_y(t) \end{bmatrix}, \quad \mathbf{q}(0) = \mathbf{q}_{\text{start}}, \quad \mathbf{q}(T) = \mathbf{q}_{\text{goal}}, \tag{1.1b}$$

$$\dot{\mathbf{q}}(0) = \begin{bmatrix} v_{0,x} \\ v_{0,y} \end{bmatrix}, \quad \dot{\mathbf{q}}(T) = \begin{bmatrix} 0 \\ 0 \end{bmatrix}, \quad \ddot{\mathbf{q}}(0) = \begin{bmatrix} a_{0,x} \\ a_{0,y} \end{bmatrix}, \quad \ddot{\mathbf{q}}(T) = \begin{bmatrix} 0 \\ 0 \end{bmatrix}, \tag{1.1c}$$

$$-\mathbf{v}_{\max} \le \dot{\mathbf{q}}(t) \le \mathbf{v}_{\max}, \quad -\mathbf{a}_{\max} \le \ddot{\mathbf{q}}(t) \le \mathbf{a}_{\max}, \quad -\mathbf{j}_{\max} \le \dddot{\mathbf{q}}(t) \le \mathbf{j}_{\max}, \tag{1.1d}$$

$$O(\mathbf{q}(t)) \cap O_i = \varnothing, \quad \forall\, i = 1, ..., N_{\text{o}}, \quad \forall\, t \in [0 \quad T], \tag{1.1e}$$

where

- $q_x(t)$ and $q_y(t)$ are respectively x and y coordinates. The start and goal positions are denoted as $\mathbf{q}_{\text{start}}$ and $\mathbf{q}_{\text{goal}}$. In this book, trajectory generation for two-dimensional robots is considered. However, the algorithm can be extended and applied to other robotic systems with higher dimensional configuration spaces.

- $\dot{\mathbf{q}}(0) = \begin{bmatrix} v_{0,x} & v_{0,y} \end{bmatrix}^T$ and $\ddot{\mathbf{q}}(0) = \begin{bmatrix} a_{0,x} & a_{0,y} \end{bmatrix}^T$ are the initial velocity and acceleration. Moreover, robots are required to be stopped at the goal postion, which leads to $\dot{\mathbf{q}}(T) = \mathbf{0}$ and $\ddot{\mathbf{q}}(T) = \mathbf{0}$.

- The velocity, acceleration and jerk are bounded by $\mathbf{v}_{max}$, $\mathbf{a}_{max}$ and $\mathbf{j}_{max}$, which correspondingly determine kinematic constraints .

- $O(\mathbf{q}(t))$ is the robot occupied area at the position $\mathbf{q}(t)$ and $\mathbf{O} = \{O_1, ..., O_i, ..., O_{N_o}\}$ is a list of obstacles. Additionally, each obstacle O_i is assumed to be a convex polygon. Nevertheless, this assumption is very universal, since general geometrical objects can be approximated by a set of convex polygons.

Solving this optimal control problem is extremely challenging, which usually requires intractable computational burden. However, an efficient algorithm solving the optimal control problem (1.1) has a huge industrial application potential. Therefore, there is a substantial motivation to develop a novel trajectory generation algorithm, which is able to outperform and mitigate some drawbacks of the state-of-the-art algorithms.

Problem 2: Centralized coordination Problem

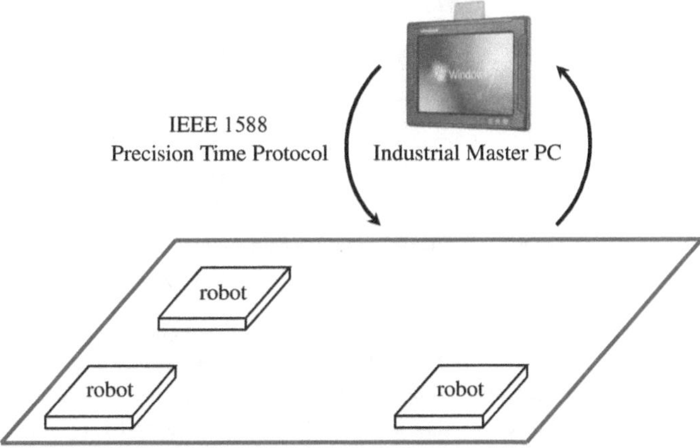

Figure 1.4: The centralized system architecture

Given a number of robots, our ultimate objective is to plan trajectories, which bring each robot from a start position to a goal postion. For several industrial applications, where the number of robots is small, all decisions can be made centrally by a master computer. As illustrated in Figure 1.4, the robots and the master computer clocks are communicated over a network, which is synchronized by standard technologies such as IEEE 1588 precision time protocol [2]. This system architecture is prevalent in industry and can be found in several products available in markets [8]. This system architecture and the objective lead to a second problem, how central control principles can be designed, which satisfies all subsequent requirements. The most important requirement is that collisions among robots as well as with static obstacles must be avoided. Additionally, deadlocks must be prevented such that each robot can reach its goal. These control principles, referred a centralized motion coordination solution, are performed by the master computer. Furthermore, these control principles should consider the solutions of **Problem 1** as inputs and operate as a subsequent process. As a consequence, the previous development results can be utilized and a modular solution is achieved.

Problem 3: Distributed Coordination Problem

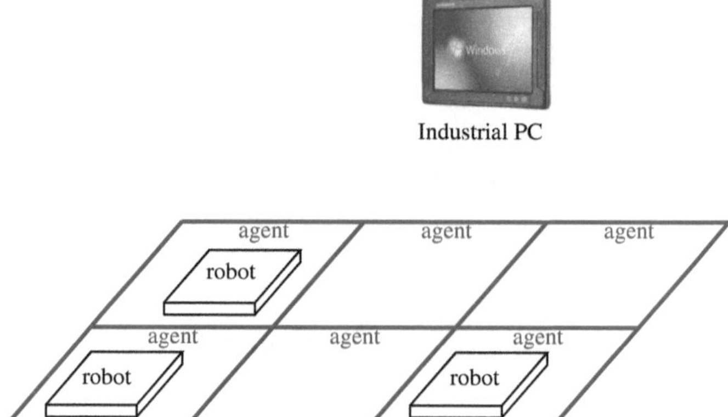

Figure 1.5: The distributed system architecture

In contrast to **Problem 2**, the system is divided into modules, namely agents. These agents are placed underneath passive robots, which actually transport wares as shown in Figure 1.5. The distributed architecture can model several industrial systems such as the aforementioned cognitive conveyor, etc. [28, 50, 87]. Moreover, decisions are not centrally taken by a master computer. Indeed, each agent decides movements for the passive robots placed on itself independently based on its local information. Due to a lack of global information and resources, the problem of designing a distributed coordination solution is more challenging. Nevertheless, applying the distributed coordination coordination solution will increase overall robustness, since the system can continue to operate in spite of the fact that a single agent is defected. Furthermore, one is able to scale up the system with the distributed coordination solution. Therefore, these articulated reasons inspire us to establish backgrounds and develop an initial solution to the distributed coordination problem. In order to design the distributed coordination solution, the following basic questions will be addressed in this work. Firstly, which information is required to be exchanged among agents? Secondly, how is the transferred information proceeded and fused together with the local information? Thirdly, how can mutual collision avoidance among the passive robots be guaranteed? In analogy to the centralized coordination solution, the solution of the **Problem 1** is employed as inputs and the distributed coordination solution operates only as a post process in order to utilize the previous development results.

1.3 book Outline

The book is organized in seven chapters. The outline is sketched in Figure 1.6. After formulating the problems, a solution architecture is introduced in Chapter 2. Moreover, fundamental backgrounds as well as a literature review about the state-of-the-art algorithms are provided in this chapter, which are relevant for understanding the book. Additionally, the scientific contributions are highlighted.

Chapter 3 explains working principles of geometrical planners, which are essential elements of the trajectory optimization algorithm. Furthermore, computational efficiency and optimality performances among the distinct geometrical planners are compared. Moreover, advantages and shortcomings are analyzed in detail. In Chapter 4, the near time-optimal trajectory generation algorithm is presented as a solution to **Problem 1**. Remarkable properties and efficiency of the algorithm are illustrated by numerical examples.

In Chapter 5, the solution to the centralized coordination **Problem 2** is introduced. Ingredient algorithms for the centralized solution are disclosed. Moreover, a theoretical invariant safety guarantee proof is given. Chapter 6 proposes an asynchronous distributed collision avoidance algorithm, which is an important component to the solution of the distributed coordination **Problem 3**. Analogously, theoretical safety guarantee can be also proven. The applicability of the centralized and distributed coordination solutions is demonstrated and verified in simulations.

In Chapter 7, a summary of the book and an outlook on future works are given.

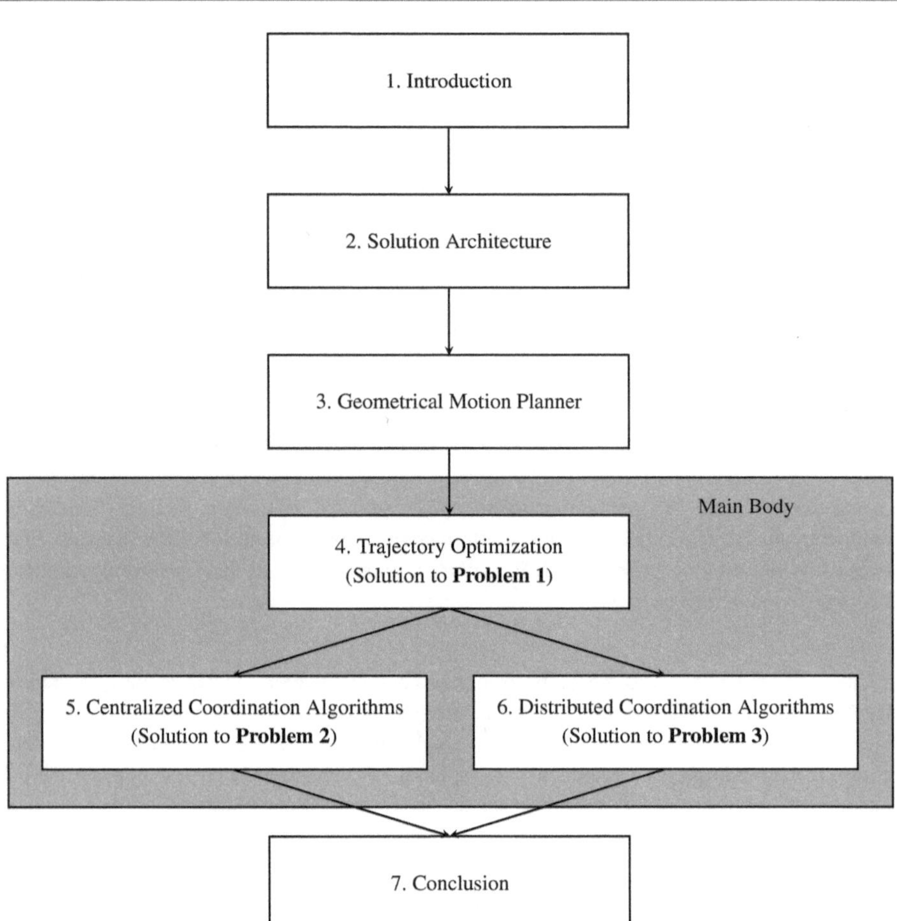

Figure 1.6: The book structure

2 Solution Architecture for Multiple Robots Motion Planning

„If you procrastinate when faced with a big difficult problem, break the problem into parts, and handle one part at a time."

Robert Collier

This statement is applied as an essential philosophy to tackle multirobot motion planning problems. The multirobot motion planning problems have been the most difficult since the first robot appearance [57]. Generally, the multirobot motion planning problems in two-dimensional configuration spaces can be briefly formulated in the following. Given two or more robots, an ultimate objective is bringing each robot from an initial position to a goal position simultaneously. Additionally, mutual collision avoidance must be ensured. Overall, performance optimality should be incorporated into these motion planning problems. An intrinsic approach is the global centralized planning, in which the robots are accumulated as a single composite robot. Therefore, a multirobot motion planning problem can be considered as a classical single robot motion planning problem in a composite joint configuration, which can be solved by applying planning algorithms for high-dimensional spaces. However, degrees of freedom of the single composite robot are not bounded and grow linearly with robot numbers. Consequently, these planning algorithms on the single composite robot has high computational burden [3, 57]. According to the best of our knowledge, no existing optimal algorithm is able to solve the multirobot motion planning problems in polynomial time. Therefore, optimality and completeness [1] are relaxed, which leads to decoupled planning approaches with remarkable low computational complexity. Based on the decoupled approaches and inspired by the aforementioned philosophy, a solution architecture for the multirobot motion planning problems is introduced in this chapter. Particularly, the solution architecture is divided into several modules, which have distinct tasks and functions. This chapter is structured as follows:

- **Background and Solution Architecture:** An overview on the decoupled planning approaches is given. Subsequently, the solution architecture is revealed.

- **Literature Review:** A literature review is provided, which gives an overview on the existing state-of-the art algorithms for each module as well as the latest development in the multirobot motion planning.

- book **Contribution:** The scientific contributions of this book are highlighted.

---

[1]*An algorithm is complete if it is guaranteed to find a solution when there is one [84, p. 80].*

2.1 Background and Solution Architecture

In order to reduce the computational complexity, the decoupled planning approaches are applied, which are presented in several publications [6, 36, 57, 70, 102]. In the decoupled planning approaches, trajectories are generated more independently. Consequently, these approaches induce low computational complexity in exchange for a loss of completeness. In other words, the approaches can fail to find an existing solution. Generally, the decoupled planning approaches can be classified into path coordination and prioritized planning [6, 36, 70, 102], which are briefly explained in the following.

Path coordination is the most prevalent decoupled planning approach [36, 70]. The idea of the path coordination is straight forward. Let's define a path for a robot i as a mapping function

$$\mathcal{P}^i(u^i) : u^i \in \begin{bmatrix} 0 & 1 \end{bmatrix} \mapsto \mathbf{q}^i, \qquad \mathcal{P}^i(u = 0) = \mathbf{q}^i_{\text{start}} \quad \text{and} \quad \mathcal{P}^i(u = 1) = \mathbf{q}^i_{\text{goal}},$$

where u^i is the path parameter and $\mathbf{q}^i$ is the robot i position. In the path coordination, the path $\mathcal{P}^i(u^i)$ for the robot i is firstly planned regardless of other robots. In order to obtain a trajectory, another mapping function from a time variable t to the path parameter u^i

$$u^i(t) : t \in \begin{bmatrix} 0 & T^i_{\text{end}} \end{bmatrix} \mapsto u \in \begin{bmatrix} 0 & 1 \end{bmatrix}$$

is required, where T^i_{end} is the total travel time of the robot i. Afterwards, scheduling algorithms are applied to avoid mutual collisions among robots [36, 70]. Nevertheless, the path coordination can lead to deadlocks. In fact, if a robot obstructs another robot from reaching its goal, the path coordination is not capable to resolve these conflicts. Therefore, another solution is derived, which is prioritized planning.

In the prioritized planning, a feasible priority sequence $\mathbf{P} = \{p_1, p_2, ..., p_{N_r}\}$ is searched and optimized [6, 102]. Let assume that a robot i has a higher priority than a robot j, then the robot j must consider the robot i as a dynamical obstacle. Nevertheless, the main difficulty of the prioritized planning is that trajectories must be generated in a time-space configuration $\mathcal{CT}$, which is complex and usually requires high computational burden [31].

Based on both the prioritized planning and the path coordination, a solution architecture is developed, which is able to combine and utilize the advantages of the both approaches. Regardless of the centralized and distributed system architecture mentioned in Chapter 1, the solution architecture consists of three fundamental components. As shown in Figure 2.1, these components are subsequently explained in the following.

- **Trajectory Optimization Algorithm:**
In the solution architecture, the trajectory optimization algorithm is a key enabler, which generates

trajectories for each robot from an initial state to a goal state. In order to guarantee that the obtained trajectories are safe, the optimization algorithm must exchange information and cooperate with a mutual collision avoidance algorithm. Additionally, the trajectory optimization algorithm must collaborate with a conflict resolution algorithm for purpose of preventing deadlocks.

- **Mutual Collision Avoidance Algorithm:**
This algorithm is employed to avoid mutual collisions among robots. In the path coordination, the scheduling algorithm can be interpreted as a collision avoidance algorithm. As aforementioned, the collision avoidance algorithm must work with the trajectory optimization algorithm.

- **Conflict Resolution Algorithm**
Finally, a conflict resolution algorithm is required in order to avoid deadlocks. Based on priority assignments, a conflict resolution algorithm is presented in Chapter 5. In this conflict resolution algorithm, a robot can be commissioned to a temporary goal in order to reserve shared free working spaces for other robots. In order to give the temporary goal to the robot, the conflict resolution algorithm must consequently communicate with the trajectory optimization algorithm as shown in Figure 2.1.

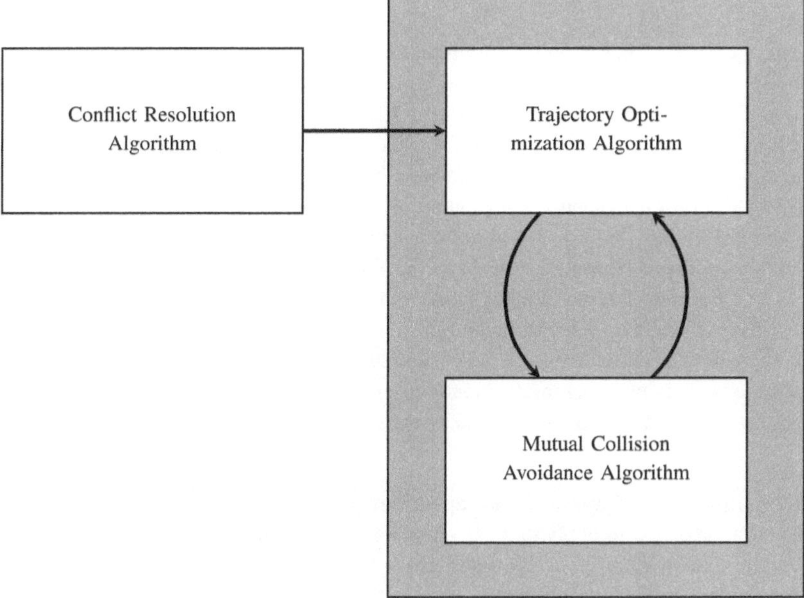

Figure 2.1: The solution architecture

2.2 Literature Review

In order to distinguish the scientific contributions, a literature review on existing algorithms for trajectory optimization, mutual collision avoidance as well as conflict resolution is given in the following. Subsequently, novel results are indicated, which are able to mitigate some disadvantages of the cutting edge algorithms.

2.2.1 Time-optimal Trajectory Generation Algorithm

As mentioned in Chapter 1, a time-optimal trajectory generation algorithm is a key to utilize robot capability. Nevertheless, time-optimal trajectory planning problems are the most challenging in robotics. Particularly, the problems become extremely complex if high-order time derivatives or kinematic constraints are considered [17, 23, 45, 46, 60, 73, 80]. Consequently, a trade-off between trajectory optimality and computational burden is introduced in practice. For solving the time-optimal trajectory planning problems, several approaches have been proposed by the control and computer science community in the last decades. In the following, these approaches as well as related works are revisited.

Trajectory generation based on velocity planning
The first approach employs geometrical path planners such as A^*, D^*, probabilistic roadmaps (PRMs) or rapid exploring random trees (RRTs). Afterwards, a list of way points is obtained. Trajectories are determined by iteratively planning velocity to a next way point. Moreover, the trajectories are smoothed by trying to delete a number of the way points during execution. Furthermore, static obstacle collision avoidance is guaranteed by incorporating *reactive methods* like Dynamic Window Approach (DWA) [30]. Due to its simplicity and easy implementation, this approach has gained a huge success. In fact, it is deployed in industrial mobile robots like Kiva System and applied by several teams in world robot soccer competitions [11, 12, 78, 110]. Nevertheless, trajectory generation based on velocity planning has two main drawbacks. Firstly, predicting ultimate smooth trajectories is difficult and overall performance such as total travel time is not optimized. Secondly, the velocity planning results in infinite jerks, which has a consequence of large tracking errors for high performance motion control systems [54].

Optimal sampling-based motion planning on state spaces
The second approach is to apply optimal sampling methods such as RRT^* on state spaces [45, 46]. Nevertheless, the main difficulty associated with this approach is computational burden due to the following reasons. Firstly, dimensions of the state spaces are at least twice that of the configuration spaces. In the problem (1.1), the state space dimension is six, which is three times as many as the configuration space dimension. Moreover, searching on high-dimensional spaces is very time consuming. Eventually, RRT^* requires optimal trajectory segments connecting any two

nodes in the state spaces, which is obtained by solving two-point boundary value problems (TB-VPs). Under consideration of state constraints, solving a TBVP is very challenging and a high computational burden is inevitable. For instance, a variant algorithm of RRT* minimizing control effort and travel time is presented by Dustin Webb [108], which is not able to constrain velocities.

Motion planning based on optimization and positive invariant sets
A natural idea corresponding to model predictive control (MPC) is employing numerical optimization for trajectory generation. Motion planning algorithms based on receding horizon optimization are introduced by Tom Schouwenaars and Yoshiaki Kuwata [52, 85]. Nevertheless, incorporating collision avoidance constraints leads to a mixed integer linear programming problem (MILP). Moreover, solving a MILP requires high computational effort. An alternative algorithm is presented by Claus Danielson [20], in which state and input constraints are handled by designing a graph of local controllers using positive invariant sets. Afterwards, a graph search algorithm is applied to find a sequence of the local controllers. A main disadvantage of this approach is that no metric exists for weighting edges of the local controller graph in order to provide an overall good performance. Moreover, computing the graph of the local controllers is also time-consuming.

The decompositional approach
The trajectory generation algorithm presented in this book is categorized as a decompositional approach, which is the most favorable and effective in practice [39, 73, 91, 93, 111]. Particularly, the near time-optimal trajectory generation is decoupled into two separated phases. First, a feasible collision-free path is generated based on geometrical way points and spline function parametrization [39, 93, 111]. Afterwards, an optimal trajectory along the specified path is planned. An ultimate trajectory is obtained by optimizing path parameters. Nevertheless, a near time-optimal trajectory along a given path can be calculated by using forward and backward integrations along with analytically solving optimal switching points. Consequently, infinite jerk profiles are induced by these proposed approaches [7, 73, 88, 91]. Moreover, the time discretization effect is not considered by these approaches.

In this book, a near time-optimal trajectory generation algorithm is presented Chapter 4, which is able to handle the drawbacks of all aforementioned approaches. For high-performance precision motion control systems, considering the time discretization effect is the most decisive determinant [54]. The proposed algorithm is able to take the time discretization effect into account. In fact, the jerk $\dddot{q}_\xi(t)$, $\xi \in \{x, y\}$, is only discontinuous at time-stamps t_d, which are integer numbers of the sampling time T_s, i.e. $t_d = nT_s$ and $n \in \mathbb{N}$. Another novel algorithm is also submitted and accepted by the International Journal of Control, Automation and Systems [76], which is not reported in the book. Based on a clever geometrical construction, this algorithm utilizes some important properties of non-uniform B-splines to smooth geometrical paths. By employing an interior point optimizer (Ipopt) [15, 105], the algorithm is capable of exploiting kinematic constraints. Consequently, this algorithm is able to generate better trajectories within much smaller computational time amounts than the state of the art algorithm [45].

2.2.2 Mutual Collision Avoidance Algorithms

In general, mutual collision avoidance algorithms can be applied to not only robotics but also other fields such as air traffic control or studying biological flocks [82, 112]. Consequently, several research results on the mutual collision avoidance algorithms have been conducted [11, 12, 78, 86, 101]. Obviously, these algorithms must be compatible with the system architectures. In other words, an appropriate algorithm can be chosen based on a predefined system architecture. As described in Chapter 1, two system architectures are considered. Specifically, the mutual collision avoidance algorithms are implemented in either the central master computer or the distributed agents. Moreover, an invariant collision-free condition must be guaranteed all the time. Due to these requirements, improper approaches can be neglected. In the following, important approaches related to our centralized and distributed collision avoidance algorithms are reviewed. Moreover, the novelty of the proposed collision avoidance algorithms is explained.

Centralized collision avoidance algorithms

An instinctive centralized approach is employing the model predictive control idea. Particularly, the trajectory optimization and the mutual collision avoidance algorithms can be formulated together as a large mixed integer programming problem (MILP) [86]. This formulation yields a large non-convex optimization problem, which can consequently be applied for a small robot number [85]. Moreover, another disadvantage of this approach is non-modularity. In contrast to the MILP formulation, dynamical safety search (DSS) is able to operate modularly as a post process. Additionally, the DSS can be deployed to a larger robot number due to its polynomial computational complexity. Nevertheless, this approach can only be applied to lower-order trajectories [11], while the proposed trajectory optimization algorithm generates high-oder trajectories. Therefore, the DSS can not be incorporated into the trajectory optimization algorithm. Moreover, stochastic performance of the DSS is not appropriate in some industrial applications. Therefore, a novel collision avoidance algorithm will be proposed in Chapter 5, which can mitigate these disadvantages of the DSS as well as be integrated with the trajectory optimization algorithm.

Distributed collision avoidance algorithms

A well-known distributed approach is the reciprocal n-body collision avoidance algorithm presented by Jur van den Berg [101]. However, it is assumed that each robot is able to sense positions and velocities of other robots. Afterwards, optimal velocities can be planned by using linear programming. Due to the additional sensor requirement, this approach can be discarded. In analogy to the centralized approaches, the model predictive control idea can be applied for solving the distributed collision avoidance problems. Indeed, a distributed MILP formulation is introduced by Yoshiaki Kuwata, which has been applied to multiple unmanned aerial vehicles [52]. By exchanging planning trajectories, a smaller MILP can be formulated and deployed in a distributed manner. Furthermore, the collision-free condition can be invariantly guaranteed by incorporating additional constraints into the distributed MILP formulation. Alternatively, a distributed collision avoidance approach based on a reserved and requested concept was proposed by Oliver Purwin [78]. In analogy to the distributed MILP formulation, collision avoidance can be theoretically guaranteed by exchanging plans and intentions among robots. Moreover, this approach is robust against stochastic communication delays. Nevertheless, the reserved and requested approach and the distributed MILP formulation can not be applied to solve our distributed collision avoidance problem due to the intrinsic distributed system architecture presented Chapter 1. Particularly, the algorithms are deployed into mobile robots instead of the static agents. Therefore, a novel distributed collision avoidance algorithm is developed, which is similarly based on the reserved and requested concept. Furthermore, the proposed distributed collision avoidance algorithm is able to operate in an asynchronous communication network. This distributed collision avoidance algorithm is presented in Chapter 6.

2.2.3 Conflict Resolution Algorithms

The collision avoidance algorithms guarantee safe movements among robots. Nevertheless, there are situations, where robots wait for others indefinitely. An example is that a robot at its goal position obstructs other robots from reaching their individual goals. These situations are defined as deadlocks, which can be resolved by deploying conflict resolution algorithms. In analogy to the collision avoidance algorithms, these conflict resolution algorithms can be categorized into centralized and distributed types, which will be explained in the following.

Centralized conflict resolution algorithms
In analogy to the centralized collision avoidance algorithms, a large MILP formulation can be employed to resolve deadlocks among multiple robots, which has intractable computational burden and consequently is impractical [52]. Contrarily, the prioritized planning is more appropriate, since it has lower computational burden than the MILP formulation. Nevertheless, the prioritized planning requires precise complex trajectory generation in a time-space configuration. In order to facilitate this trajectory generation, the time-space configuration is approximately split into a grid. Subsequently, an optimal priority sequence can be found by applying the A^* and hill-climbing algorithms. In Chapter 5, this approach will be presented .

Distributed conflict resolution algorithms
In the distributed conflict resolution algorithms, considerable results are usually based on a set of rules [34, 95]. Particularly, a behavior-based model is proposed by Dali Sun [95]. Under an assumption that a robot is able to sense critical front areas, a set of traffic rules can be designed. This algorithm is robust and was demonstrated during an industrial fair in Stuttgart [95]. However, the proposed algorithm is not able to resolve conflicts involving more than two robots. Consequently, deadlock-free systems can not be guaranteed. Contrarily, a deadlock-free algorithm is proposed by Kevin Gue for a system, which consists of rectangular agents and is similar to the aforementioned distributed system architecture [34]. This algorithm can be briefly summarized in the following. Each agent has two possible states, which are empty and occupied. Based on these states and intentions of neighboring agents, a set of rules can be constructed, which guarantees a deadlock-free system. Nevertheless, robot start and goal locations are limited by this algorithm. Indeed, all

robots are assumed to start at one side and end at another side of conveyers. Consequently, the algorithm can not be deployed to any desired conveyor architecture. Furthermore, full flexibility as well as aggressive maneuvers for the robots can not be achieved by this algorithm, since only four cardinal primitive motions are allowed. Although the algorithm can be employed directly to our system, there is nonetheless a great demand for distributed conflict resolution algorithms in general cases.

3 Geometrical Motion Planner

A geometrical motion planer is an essential element of the solution architecture. In this chapter, a thorough overview on the geometrical motion planners is provided. Moreover, rapid exploring random tree (RRT) and A* grid-based geometrical planners are chosen due to their efficiency. Subsequently, these geometrical planners are described in detail. Finally, performances and applicability of these geometrical planners are compared and discussed.

3.1 On Existing Approaches to Geometrical Motion planners

Geometrical motion planning has been well studied for the last decades [57, 58]. The objective of geometrical motion planners is to find a set of way points $\mathbf{W} = \left\{ \mathbf{w}_1 := \mathbf{q}_{\text{start}}, \mathbf{w}_2, ..., \mathbf{w}_{N_\text{w}} := \mathbf{q}_{\text{goal}} \right\}$ from a start position to a goal position, where every straight line $\mathbf{w}_i \mathbf{w}_{i+1}$ connecting any two consecutive way points is collision-free. Interested readers are referred to several surveys on geometrical motion planning [11, 57, 58]. According to the best of our knowledge, no geometrical motion planner exists, which guarantees to return a path when one exists. In practice, approximate algorithms are consequently applied, which are either resolution or probabilistic complete. In the following, these approximate algorithms are introduced. Subsequently, the well-established theoretical properties as well as the basic working principle of these geometrical motion planners are respectively explained.

• **Sampling based approach**

The randomized sampling based planning is the most modern approach, which has attracted a great attention from industrial as well as academical communities [48]. Due to its rapid exploration, this approach is suitable for solving the robot geometrical planning problem in high-dimensional configuration spaces. Particularly, randomized sampling geometrical planners are mainly divided into two classes, which are probabilistic roadmap (PRM) and rapid exploring random tree (RRT) [41, 42, 44, 47, 51, 59]. PRM is appropriate in slow changing environments. In PRM, the geometrical motion planning is separated into two stages, which are a learning phase and a query phase [32, 47]. In the learning phase, a random graph $\mathcal{G}$ of configuration spaces is generated. By applying graph search methods such as A* or Dijkstra algorithm, a geometrical path can be found in the query phase. In contrast to PRM, RRT is suitable in fast changing environments. Particularly, a random tree $\mathcal{T}$ is built online in a RRT algorithm [59]. Afterwards, a simple tree search algorithm can be applied in order to find a geometrical path. In the last decades, several variant and extended RRT algorithms have been developed [11, 51]. Moreover, the working environment for our multirobot systems can change dynamically and fast. Consequently, the RRT algorithm is chosen as one of the geometrical planners in our implementation, which will be introduced in the following section.

• Grid-based search approach

Another common approach for the geometrical path planning is based on classical grid search. In spite of complex environment topologies, the configuration space can be represented by an occupancy grid of rectangular cells. Based on the occupancy grid, the geometrical motion planning problem can be solved by using search methods such as Dijkstra, A^* or D^* algorithms [69, 94]. Among them, A^* algorithm is the most popular and well-known. Moreover, speed of the geometrical motion planner can be significantly improved by incorporating simple guiding heuristics into the A^* algorithm. Consequently, an A^* grid-based geometrical planner is implemented as an alternative solution, which will be presented in this chapter.

Finally, the rest of the chapter is structured by the following sections:

- **Rapid Exploring Random Tree Geometrical Planner:** The working principle of rapid exploring random tree (RRT) is explained. Moreover, functional primitives of RRT are described in detail. Subsequently, a collision detection for RRT is also provided. In order to optimize the geometrical path, a line simplification algorithm is introduced, which reduces and minimizes the way point number.

- **An A^* Grid-based Geometrical Planner:** The A^* algorithm is described. Subsequently, the A^* grid-based geometrical planner is presented.

- **Performance Comparison:** Performance and applicability of the RRT and A^* grid-based geometrical planners are compared in various scenarios.

- **Conclusion:** Eventually, some remarks are concluded in this chapter.

3.2   Rapid Exploring Random Tree Geometrical Planner

RRT has been widely applied for autonomous robots, which is suitable for solving motion planning problems in a fast changing environment. In fact, RRT offers an efficient way to obtain the geometrical path. Furthermore, extended RRT algorithms can be applied to solve more difficult problems such as kinodynamical and nonholonomic motion planning [42, 60]. However, these extended algorithms often require high computation burden, which is not appropriate for industrial applications. In this book, a simple RRT geometrical planner is considered, which has attained a tremendous success from world robot soccer competitions to industrial applications [11, 43, 56].

In the following, a brief introduction of RRT is given. Subsequently, all ingredients of the RRT geometrical planner are provided.

3.2.1 The Basic Working Principle of RRT

Algorithm 3.1 planRRT($\mathbf{q}_{start}$, $\mathbf{q}_{goal}$, $\mathbf{O}$)

    Input: start and goal configurations $\mathbf{q}_{start}$, $\mathbf{q}_{goal}$, and a list of obstacle $\mathbf{O}$
    Output: a constructed tree $\mathcal{T}$ and a list of way point $\mathbf{W}_i$
1: $\mathcal{T}$.root $\leftarrow$ $\mathbf{q}_{start}$;
2: **do**
3:     $\mathbf{q}_{rand}$ $\leftarrow$ randomConfiguration(p_{goal}, $\mathbf{q}_{goal}$);
4:     $\mathbf{q}_{nearest}$ $\leftarrow$ nearestConfiguration($\mathbf{q}_{rand}$, $\mathcal{T}$);
5:     $\mathcal{T}$ extendTree($\mathcal{T}$, $\mathbf{q}_{rand}$, $\mathbf{q}_{nearest}$, $\mathbf{O}$);
6: **while** ($\mathbf{q}_{goal} \notin \mathcal{T}$)
7: $\tilde{\mathbf{W}}$ constructGeometricalPathFromTree($\mathcal{T}$, $\mathbf{q}_{goal}$)

RRT was firstly introduced by Steven LaValle and James Kuffner [59] and is an important breakthrough in robot motion planning. Thereafter, extensive researches on the RRT have been conducted [11, 51]. Nevertheless, the basic working principle of the RRT is surprisingly simple, which is summarized as pseudocode given in Algorithm 3.1. Basically, a tree $\mathcal{T}$ is rooted at the start configuration $\mathbf{q}_{start}$ and iteratively constructed by adding new random configurations. In each iteration, a random configuration $\mathbf{q}_{rand}$ is generated by a *randomConfiguration*(p_{goal}, $\mathbf{q}_{goal}$) function. Subsequently, a nearest configuration $\mathbf{q}_{nearest}$ from the tree $\mathcal{T}$ to the random configuration $\mathbf{q}_{rand}$ is selected by a *nearestConfiguration*($\mathbf{q}_{rand}$, $\mathcal{T}$) function. Lastly, the tree $\mathcal{T}$ is extended by employing a *extendTree*($\mathcal{T}$, $\mathbf{q}_{rand}$, $\mathbf{q}_{nearest}$) function. In the following, these functions are described in detail.

- **Random configuration function:**
In each iteration, a random configuration $\mathbf{q}_{rand}$ is chosen. For practical applications, a heuristic is introduced by James Bruce [11]. In fact, a probability p_{goal} for picking the goal configuration $\mathbf{q}_{goal}$ is integrated into the function in order to guide the search towards the goal as shown in lines 2-3 of Algorithm 3.2. Subsequently, the tree grows biasedly towards the goal area. Moreover, the larger the probability p_{goal} is, the greedier the RRT planner is. Nevertheless, the larger the probability p_{goal} is, the less configuration space will be explored. Consequently, it leads to a higher probability of not finding a path. Therefore, there is a trade-off in choosing the probability p_{goal}. Several empirical research results have been conducted for finding an optimal probability p_{goal} [11]. Obviously, the optimal probability p_{goal} depends on environmental topologies and applications. For example, the optimal probability lies in the interval between 0.1 and 0.3 for the holonomic robot motion planning in world soccer competitions [11].

Algorithm 3.2 randomConfiguration(p_{goal}, q_{goal})

1: var $p \in \mathbb{R}$;
2: $p \leftarrow$ uniformRandom$(0, 1)$;
3: **if** $(p \leq p_{goal})$ **then**
4:     $q_{rand} \leftarrow q_{goal}$;
5: **else**
6:     $q_{rand} \leftarrow$ getARandomConfiguration();
7: **end if**

- **Nearest configuration function:**

In this function, a configuration $q_{nearest}$ from the tree $\mathcal{T}$ is chosen such that the distance from the configuration $q_{nearest}$ to the random configuration q_{rand} is minimal as shown in lines 2-5 of Algorithm 3.3. For the RRT geometrical planner, the distance is measured as Euclidean norm. If $N_d = |\mathcal{T}|$ is defined as the number of tree configurations, search complexity for finding the nearest configuration is $\mathcal{O}(N_d)$. Nonetheless, if k-d tree data structure is used, the search complexity can be reduced to $\mathcal{O}(log\,N_d)$ on average [11]. However, the primitive search is applied for the sake of simplifying implementation.

Algorithm 3.3 nearestConfiguration(q_{rand}, $\mathcal{T}$)

    Input: a random configuration q_{rand} and a tree $\mathcal{T}$
    Output: a nearest configuration $q_{nearest}$ of the tree to the random configuration
1: **initialize** $q_{nearest}$;
2: **for each** $q_{node} \in \mathcal{T}$ **do**
3:     **if** distance$(q_{node}, q_{rand}) <$ distance$(q_{nearest}, q_{rand})$ **then**
4:        $q_{nearest} \leftarrow q_{node}$;
5:     **end if**
6: **end for**

- **Extend tree function:**

The objective of this function is to incorporate the new random configuration q_{rand} into the existing tree $\mathcal{T}$. For the RRT geometrical planner, this function can be easily implemented and briefly explained as follows: First, it is checked whether the robot collides with any obstacle on the list **O** at the random configuration q_{rand}. Afterwards, it will be evaluated if the straight line connection between the nearest configuration $q_{nearest}$ and the new random configuration q_{rand} is collision-free. If the both collision-free conditions are satisfied, then the random configuration q_{rand} is integrated into the existing tree $\mathcal{T}$. Particularly, the new random configuration q_{rand} is connected as a child of the nearest configuration $q_{nearest}$. Conversely, the nearest configuration $q_{nearest}$ is considered as a parent of the new random configuration q_{rand} as stated in lines 3-4 of Algorithm 3.4. This extension procedure is additionally illustrated in Figure 3.1.

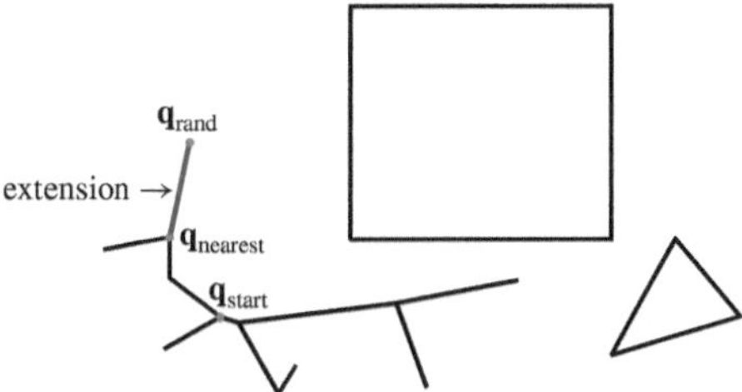

Figure 3.1: The extend tree function. After checking collisions, the tree is extended by a new configuration $\mathbf{q}_{\text{rand}}$ and a new blue edge.

Algorithm 3.4 extendTree($\mathcal{T}$, $\mathbf{q}_{\text{rand}}$, $\mathbf{q}_{\text{nearest}}$, $\mathbf{O}$)

    Input: a tree $\mathcal{T}$, a random configuration $\mathbf{q}_{\text{rand}}$, a nearest configuration $\mathbf{q}_{\text{nearest}}$, a list of obstacles $\mathbf{O}$.
    Output: an extended tree $\mathcal{T}$
1: **if** $\big(\text{collisionFree}(\mathbf{q}_{\text{rand}}, \mathbf{O}) \wedge \text{collisionFree}(\mathbf{q}_{\text{rand}}, \mathbf{q}_{\text{nearest}}, \mathbf{O})\big)$ **then**
2:     child $[\mathbf{q}_{\text{nearest}}] \leftarrow \mathbf{q}_{\text{rand}}$;
3:     parent $[\mathbf{q}_{\text{rand}}] \leftarrow \mathbf{q}_{\text{nearest}}$;
4: **end if**
5: **return** $\mathcal{T}$;

As soon as the goal configuration is found, Algorithm 3.1 will terminate. Furthermore, a rough path $\tilde{\mathbf{W}}$ can be constructed based on the building tree $\mathcal{T}$ from the goal configuration $\mathbf{q}_{\text{goal}}$ to the start configuration $\mathbf{q}_{\text{start}}$. The RRT algorithm is well-known for its efficiency in high-dimensional configuration spaces, since it relaxes the completeness and explores greedily the working space. Consequently, the RRT algorithm is only probabilistic complete. In other words, the probability of finding an existing path tends to one, as the number of iterations approaches infinity [58]. As aforementioned, the RRT algorithm is dominated by collision detection in the extend tree function. Therefore, an efficient collision detection algorithm can improve the speed of the RRT geometrical planner profoundly. Additionally, a collision detection algorithm can be adopted in the trajectory optimization proposed in Chapter 4. Consequently, a collision detection algorithm is introduced and some supplementary tricks for increasing its efficiency are presented in the following.

3.2.2   A Collision Detection Algorithm

As aforementioned, collision detection is considered as a computational bottleneck of the geometrical planner. Moreover, collision detection is an important research area in computational geometry, which is encountered in several fields such as robotics, manufacturing, computer simulation and animation [9, 16, 62, 63]. Particularly, interested readers are referred to a survey on the collision detection algorithms [63]. Moreover, shortcomings and advantages of numerous algorithms are also discussed in the survey. Generally, robot and obstacle representation models are required in order to implement a collision detection algorithm. Based on these models, an appropriate collision detection algorithm can be chosen. Therefore, an overview on these models is firstly given in the following.

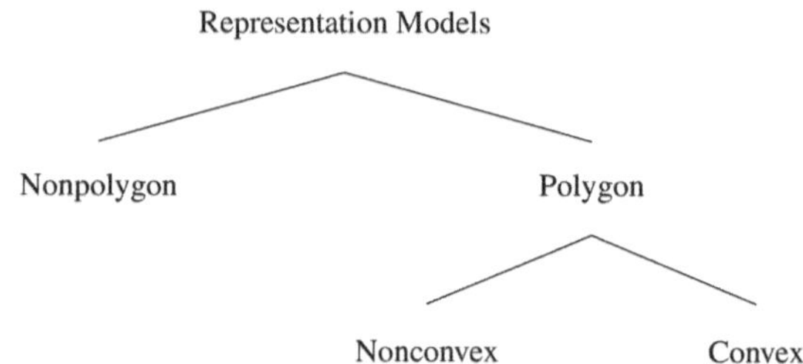

Figure 3.2: The representation model classification

As shown in Figure 3.2, the representation models can be classified into nonpolygon and polygon [63]. Furthermore, the polygon is categorized as either convex or nonconvex. If the polygon is convex, this property can be exploited to develop an efficient collision detection algorithm. Moreover, all representation models can be approximated by convex polygons [63]. Therefore, the convex polygons are chosen to represent robots as well as obstacles in this book. In order to detect collisions between two convex polygons, the hyperplane separation theorem can be applied, which is illustrated in Figure 3.3 and stated in the following.

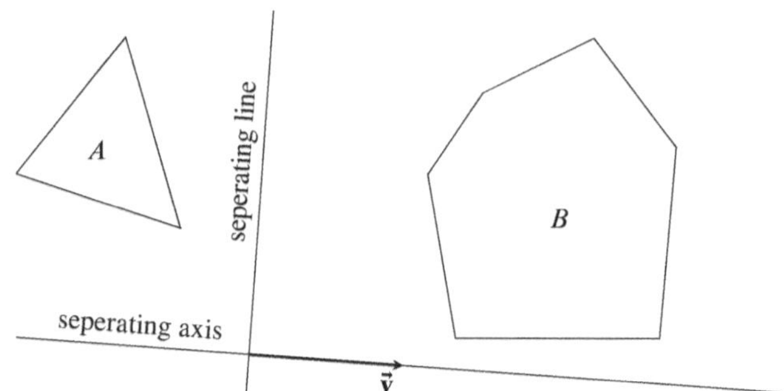

Figure 3.3: The hyperplane separation theorem

Theorem 3.1 (Hyperplane Separation Theorem [10]) *Let A and B be two disjoint convex subsets of $\mathbb{R}^n$. Then there exists a nonzero vector $\mathbf{v}$ and a real number $c \in \mathbb{R}$ such that*

$$\langle x, \mathbf{v} \rangle \geq c \quad and \quad \langle y, \mathbf{v} \rangle \leq c \tag{3.1}$$

for all x in A and y in B. In other words, the hyperplane is defined by $\langle \bullet, \mathbf{v} \rangle = c$, where $\mathbf{v}$ is the normal vector separating A and B.

Based on the hyperplane separation theorem, a primitive collision detection algorithm is implemented, whose pseudocode is given in Algorithm 3.5. The algorithm is able to identify overlapping of any two convex polygons A and B efficiently. Furthermore, the working principle of the collision detection algorithm can be briefly explained in the following. A separating line exists, if and only if two convex polygons can be divided by at least one of their edges. This condition can be verified easily by using the scalar product as stated in Theorem 3.1.

Algorithm 3.5 A collision detection algorithm

    Input: two convex polygons A and B.

    Output: return **true** if there is an overlapping between A and B and **false** otherwise.

1: test1 $\leftarrow$ isAnyEdgeOfASeparatingLine();
2: test2 $\leftarrow$ isAnyEdgeOfBSeparatingLine();
3: **if** $\big(($test1 $=$ **true**$) \vee ($test2 $=$ **true**$)\big)$ **then**
4:     collision $\leftarrow$ **false**;
5: **else**
6:     collision $\leftarrow$ **true**;
7: **end if**

If N_v is defined as the maximal vertices number of the convex polygons, the collision detection algorithm has a computational complexity of $\mathcal{O}(N_v^2)$. Moreover, if the number of obstacles is large, a hierarchical space partitioning representation can be applied [62]. In fact, collision-free cases can be quickly found out when two objects are far apart from each other. Nevertheless, the number of obstacles is small in our applications, the strategy is not applied to the collision detection algorithm. Consequently, Algorithm 3.5 is chosen in order to detect collisions for the sake of simplicity.

3.2.3   Line Simplification Algorithm

Line simplification algorithms are very crucial in vector graphics, cartography, computer aided design and robotic motion planning [24, 68, 98]. For these applications, a main objective is reducing storage spaces and increasing processing speeds by eliminating superfluous points. Due to the wide and important applications, several line simplification algorithms have been developed, which include Douglas Peucker, Reumann-Witkam, Lang or perpendicular distance line simplification [24, 55, 81, 98]. For the sake of simplicity, the perpendicular distance algorithm is chosen and modified for the geometrical motion planner, which is described in the following.

Algorithm 3.6 A line simplification algorithm

    Input: a list of initial way points $\tilde{\mathbf{W}} = \{\tilde{\mathbf{w}}_1, \tilde{\mathbf{w}}_2, ..., \tilde{\mathbf{w}}_{N_w}\}$ and a list of obstacles $\mathbf{O}$

    Output: a list of smooth way points $\mathbf{W}$

1: **initialize W**;

2: **if** $(\text{length}(\tilde{\mathbf{W}}) \leq 2)$ **then**

3:     $\mathbf{W} \leftarrow \tilde{\mathbf{W}}$;

4:     **return**;

5: **else**

6:     **if** $(\text{collisionFree}(\tilde{\mathbf{w}}_1, \tilde{\mathbf{w}}_3, \mathbf{O}))$ **then**

7:         $\tilde{\mathbf{W}}_{\text{new}} \quad \{\tilde{\mathbf{w}}_1, \tilde{\mathbf{w}}_3, ..., \tilde{\mathbf{w}}_{N_w}\}$;   *// The way point $\tilde{\mathbf{w}}_2$ is discarded.*

8:         $\mathbf{W} \leftarrow \text{lineSimplification}(\tilde{\mathbf{W}}_{\text{new}}, \mathbf{O})$;

9:     **else**

10:        $\tilde{\mathbf{W}}_{\text{new}} \quad \{\tilde{\mathbf{w}}_2, \tilde{\mathbf{w}}_3, ..., \tilde{\mathbf{w}}_{N_w}\}$;

11:        $\mathbf{W}_{\text{rest}} \quad \text{lineSimplification}(\tilde{\mathbf{W}}_{\text{new}}, \mathbf{O})\}$;

12:        $\mathbf{W} \leftarrow \{\tilde{\mathbf{w}}_1\} \cup \mathbf{W}_{\text{rest}}$;

13:     **end if**

14: **end if**

The RRT algorithm delivers a list $\tilde{\mathbf{W}} = \{\tilde{\mathbf{w}}_1, \tilde{\mathbf{w}}_2, ..., \tilde{\mathbf{w}}_{N_w}\}$, which contains unnecessary rough way points. Based on the obstacle list $\mathbf{O}$, the line simplification algorithm refines the list of way points $\tilde{\mathbf{W}}$ in order to obtain a minimal necessary way points $\mathbf{W}$. The pseudocode of the line simplification algorithm can be found in Algorithm 3.6. The idea of the line simplification algorithm can be briefly explained in the following. Firstly, an empty list of smooth way points $\mathbf{W}$ is initialized. If the number of way points is equal or less than 2, then the smooth way points $\mathbf{W}$ are assigned by the initial way points $\tilde{W}$ as shown in lines 3-4 of Algorithm 3.6. In other words, no simplification procedure is needed. Contrarily, the first three way points $\tilde{\mathbf{w}}_1$, $\tilde{\mathbf{w}}_2$ and $\tilde{\mathbf{w}}_3$ are considered. Instead of comparing the perpendicular distance from the way point $\tilde{\mathbf{w}}_2$ to the line $\tilde{\mathbf{w}}_1\tilde{\mathbf{w}}_3$ with a maximal distance threshold as the original perpendicular distance line algorithm, a line connection $\tilde{\mathbf{w}}_1\tilde{\mathbf{w}}_3$ is constructed and checked for collision with the obstacle list $\mathbf{O}$. If no collision exists, the way point $\tilde{\mathbf{w}}_2$ can be discarded from the list $\tilde{\mathbf{W}}$. Subsequently, the next three way points $(\tilde{\mathbf{w}}_1, \tilde{\mathbf{w}}_3, \tilde{\mathbf{w}}_4)$ will be continuously proceeded as aforementioned. Otherwise, if the line connection $\tilde{\mathbf{w}}_1\tilde{\mathbf{w}}_3$ collides with the obstacle list $\mathbf{O}$, then the way point $\mathbf{w}_1$ is added to the smooth way point list $\mathbf{W}$. As stated in lines 10-13 of Algorithm 3.6, a similar procedure will be applied to the remaining way points $\mathbf{W}_{\text{new}} = \{\tilde{\mathbf{w}}_2, \tilde{\mathbf{w}}_3, ..., \tilde{\mathbf{w}}_{N_w}\}$. An example of our line simplification algorithm is illustrated in Figure 3.4. Finally, the line simplification algorithm has a complexity of $\mathcal{O}(N_w^2)$, where N_w is the number of the initial way points.

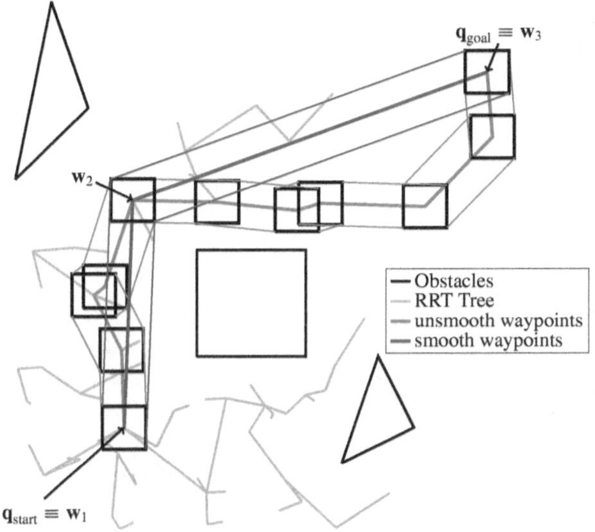

Figure 3.4: Illustration of the line simplification algorithm. The RRT is depicted by the green lines. The obstacles and the robot are shown by the black lines. Finally, the initial geometrical path $\tilde{\mathbf{W}}$ is presented by the red lines, while the refined geometrical path $\mathbf{W} = \{\mathbf{w}_1, \mathbf{w}_2, \mathbf{w}_3\}$ is demonstrated by the blue lines.

3.3　An A* Grid-based Geometrical Planner

An alternative geometrical path planner is using a tree or graph search method based on an occupancy grid map. Particularly, the environment is discretized into a grid of equally spaced cells, which are either free or occupied. If a robot at a cell center position does not collide with any obstacle, then the cell is free. Conversely, the cell is occupied. In general, several existing methods such as greedy best first search, A* or Dijkstra's algorithms are able to find a collision-free path from a start cell to a goal cell. Nevertheless, A* algorithm is the most well-known and prevalent in robotic community due to its optimality and efficiency. Therefore, A* algorithm is chosen to solve the geometrical motion planning problem. In the following, the working principle and properties of A* algorithm are introduced.

Given a start cell n_{start}, a goal cell n_{goal} and an occupancy grid map G, A* algorithm finds the shortest path by applying a search procedure, which are described in several literatures [38, 72, 84]. Basically, A* algorithm evaluates each cell n_i by three functions $g(n_i)$, $h(n_i)$ and $f(n_i)$. The function $g(n_i)$ is an exact cost from the start cell n_{start} to the cell n_i, while the function $h(n_i)$ estimates heuristically a cost from the cell n_i to the goal cell n_{goal}. Additionally, the function

$f(n_i)$ is a sum of the functions $g(n_i)$ and $h(n_i)$, which can consequently be interpreted as an estimated cost of the shortest path constrained to go through the cell n_i. Therefore, $f(n_i)$ can be used to guide the search procedure more efficiently [72]. Principally, the search procedure can be summarized by pseudocode given in Algorithm 3.7 and constituted by the following steps.

- **Step 1:** As shown in lines 1-2 of Algorithm 3.7, an empty open list is created and the start cell n_{start} is put to the open list. The open list contains all cells, which will be processed. Similarly, an empty close list is created. The close list contains all cells, which have already been explored. If the open list is empty, the search will be stopped with failure. Otherwise, go to Step 2.

- **Step 2:** Find a cell n_{min} on the open list with the minimal estimate $f(n_{\text{min}})$. Expectedly, the shortest path through the cell n_{min} has the minimal cost. In other words, n_{min} is the most promising cell on the open list. Afterwards, the cell n_{min} is removed from the open list and added into the close list. Moreover, a list of successor cells $\mathcal{N}_i$ is generated. In the geometrical planner, A^* algorithm is implemented for a two dimensional grid map. Subsequently, eight neighbors of the cell n_{min} are candidates for the successor cells. Additionally, these successor cells are required to be collision-free.

- **Step 3:** For each successor cell n_i, all three functions $f(n_i)$, $g(n_i)$ and $h(n_i)$ are respectively computed. In fact, the function $g(n_i)$ is the real cost from the start cell n_{start} to the cell n_i. Consequently, $g(n_i)$ is a sum of the cost $g(n_{\text{min}})$ from start cell n_{start} to the cell n_{min} and a cost from the cell n_{min} to the cell n_i. Subsequently, the function $h(n_i)$ is a heuristic estimate of the cost from the cell n_i to the goal cell n_{goal} without considering obstacles, which is basically the distance from the cell n_i to the cell n_{goal}. Eventually, an estimate $f(n_i)$ of an optimal path constrained to go through the cell n_i is evaluated, which is a sum of the exact cost $g(n_i)$ and the heuristic estimate $h(n_i)$ as stated in lines 13-15 of Algorithm 3.7.

- **Step 4:** If the successor cell n_i is already on either *open_list* or *close_list* and the previous estimate $\hat{f}(n_i)$ is bigger than the current estimate $f(n_i)$, then the parent of the cell n_i will be redirected to the cell n_{min} as shown in lines 16-27 of Algorithm 3.7. Contrarily, if the successor cell n_i is neither on *open_list* nor *close_list*, then it will be added to the *open_list* and go back to **Step 2**.

- **Step 5:** If the goal cell n_{goal} is found, the search procedure is stopped. As shown in lines 34-36 of Algorithm 3.7, the optimal path is finally constructed by iteratively tracing back the parent cell, until the start cell n_{start} is found.

Algorithm 3.7 A^* search algorithm

    Input: a start cell n_{start}, a goal cell n_{goal} and a grid map G
    Output: a list of node $\mathcal{L}$ from the start cell to the goal cell.

1: **initialize** *open_list* and *close_list*;
2: *open_list* $\leftarrow$ *open_list* $\cup$ $\{n_{\text{start}}\}$;   *// add the start cell to the open list*
3: **while** $(\text{isNotEmpty}(\textit{open\_list}))$ **do**

4:	$n_{\min}$ findMinimalCostNode(*open_list*);	// *find the cell with minimal estimate cost f*

4:       $n_{\min}$ findMinimalCostNode(*open_list*);   // *find the cell with minimal estimate cost f*

5:      $close\_list \leftarrow close\_list \cup \{n_{\min}\}$;   // *add the cell n_{min} to the close list*

6:      $open\_list \leftarrow open\_list \setminus \{n_{\min}\}$;   // *remove the cell n_{min} from the open list*

7:      $\mathcal{N}$ getFreeSuccessor($n_{\min}$);   // *generate a successor cell list of the cell n_{min}*

8:      **for each** $n_i \in \mathcal{N}$ **do**

9:         parent(n_i) $n_{\min}$;   // *set the n_{min} as the parent cell of the successor cell n_i*

         // *If the goal cell is found, the search procedure is stopped.*

10:         **if** ($n_i \equiv n_{\text{goal}}$) **then**

11:            stopTheSearch();

12:         **end if**

         // *Evaluate all cost for each successor cell n_i*

13:         $g(n_i)$ $g(n_{\min}) + \text{distance}(n_{\min}, n_i)$;

14:         $h(n_i)$ $\text{distance}(n_i, n_{\text{goal}})$;

15:         $f(n_i)$ $g(n_i) + h(n_i)$;

         // *If the cell n_i is on the open_list*

16:         **if** $\big((n_i \equiv n_j) \wedge (n_j \in open\_list)\big)$ **then**

17:            **if** ($f(n_i) < f(n_j)$) **then**

18:               $f(n_j)$ $f(n_i)$;

19:               parent(n_j) $n_{\min}$;   // *redirect the parent of the cell n_i*

20:            **end if**

21:         **end if**

         // *If the cell n_i is on the close_list*

22:         **if** $\big((n_i \equiv n_j) \wedge (n_j \in close\_list)\big)$ **then**

23:            **if** ($f(n_i) < f(n_j)$) **then**

24:               $f(n_j)$ $f(n_i)$;

25:               parent(n_j) $n_{\min}$;

26:            **end if**

27:         **else**

28:            $open\_list \leftarrow open\_list \cup \{n_i\}$;

29:         **end if**

30:      **end for**

31:   **end while**

32: $\mathcal{L}$ $\mathcal{L} \cup \{n_{\text{goal}}\}$;

33: n_k n_{goal};

34: **while** ($n_k \not\equiv n_{\text{start}}$) **do**

35:   n_k parent(n_k);

36:   $\mathcal{L}$ $\{n_k\} \cup \mathcal{L}$;

37: **end while**

A^* algorithm is widespread due to its profound properties [38, 72], which include completeness, optimality and computational complexity. In the following, these properties are introduced.

A^* algorithm is complete [72]. In other words, if a solution path exits, A^* algorithm is able to return the path $\mathcal{L}$. The completeness of A^* algorithm can be shortly explained. The search procedure returns a failure when the open list is empty. If a solution path $\mathcal{L}$ exists, then the open list can not be empty due to the following reason. Every cell on the existing path has at least one successor, which will be added to the open list. Consequently, there is at least one cell on the open list. Eventually, the path is explored until the goal node n_{goal} is discovered.

The condition for the optimality of A^* algorithm depends on search methods. In Algorithm 3.7, a tree-based search method is applied. Consequently, A^* algorithm delivers an optimal solution if the heuristic estimate function $h(n_i)$ is admissible [38]. Moreover, a heuristic estimate function $h(n_i)$ is admissible, if it does not overestimate a real optimal cost in the **Step** 3. In other words, the admissibility can be formulated as

$$h(n_i) \le c^*(n_i),$$

where $c^*(n_i)$ is the real optimal cost from the cell n_i to the goal cell n_{goal}. In Algorithm 3.7, the Manhattan, Euclidean, and infinity norm distances between two cells can be used as the heuristic functions, which are provided in Appendix A.2. Since obstacles are neglected, these heuristic functions are minimal and less than the real cost $c^*(n_i)$. Consequently, these heuristic functions $h(n_i)$ are admissible.

Basically, the computational complexity is measured by counting a number of elementary operators performed by an algorithm. A^* algorithm has a computational complexity of $\mathcal{O}(b^d)$, where b is the branching factor and d is the depth of the path $\mathcal{L}$. The branching factor b is the average number of the successors. In the two dimensional search space, the maximal number of successors is eight, while the depth d is the cell number of the obtained path, i.e. $d = |\mathcal{L}|$.

Based on A^* algorithm, a geometrical motion planner is developed and briefly described in the following. Given a list of obstacles $\mathbf{O}$, a start configuration $\mathbf{q}_{\text{start}}$ and a goal configuration $\mathbf{q}_{\text{goal}}$, a list of way points $\mathbf{W} = \{\mathbf{w}_1 := \mathbf{q}_{\text{start}}, \mathbf{w}_2, \ldots, \mathbf{w}_{N_w} := \mathbf{q}_{\text{goal}}\}$ can be found by using the following procedures. First, a two dimensional grid map $\mathbf{G}$ is generated, in which the occupied cells are illustrated by the red squares as shown in Figure 3.5. Subsequently, a cell n_{start} and a cell n_{goal} are determined, which are respectively the free cells nearest to the start and goal configurations. By applying A^* algorithm, a list of cells $\mathcal{L}$ can be found, which is depicted as the blue cells in Figure 3.5. Afterwards, the start and goal configurations as well as the centers of the cells on the list $\mathcal{L}$ are concatenated as a list of way points $\tilde{\mathbf{W}}$. Eventually, unnecessary way points are eliminated by the aforementioned line simplification algorithm and a geometrical path $\mathbf{W}$ is obtained. Nevertheless, one is not always able to find an existing solution by applying the A^* grid-based geometrical planner due to the discretized cell size. Consequently, the A^* grid-based geometrical planner is only resolution complete in contrast the probabilistic completeness of the RRT geometrical planner. In the following section, performance and efficiency of both geometrical planners are demonstrated and compared.

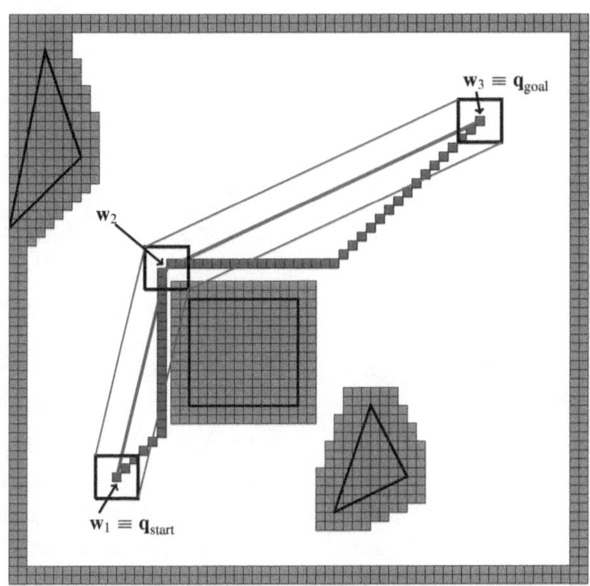

Figure 3.5: An example of the A* grid-based geometrical motion planner. The blue geometrical path $\mathbf{W} = \{\mathbf{w}_1, \mathbf{w}_2, \mathbf{w}_3\}$ is obtained by by applying the A* grid-based geometrical motion planner.

Table 3.1: The parameter values for RRT and A* grid-based geometrical planner

RRT parameter	Value	A* parameter	Value
Target probability p_{goal}	0.1	Cell size	0.025 m
Maximum allowable iteration number	1750	Heuristic cost	Euclidean Distance

3.4 Performance Comparison

In this book, the geometrical motion planners are developed for robots operating in a two-dimensional space. Consequently, it is reasonable to analyze and compare these geometrical motion planners in several two-dimensional scenarios. These scenarios are depicted in Figure 3.6, which were used for evaluating distinct motion planning algorithms [11]. In order to generalize results, a broad range of space connectivity is covered by these scenarios. In fact, the numbers of obstacles vary from 0 to 8. Since the RRT geometrical planner delivers stochastic results, performances are measured by mean values of 1000 executions. The RRT and A* parameters are provided in Table 3.1. Furthermore, all measurements are done in MATLAB$^{\circledR}$ 2017b environment running on a Intel$^{\circledR}$ Core^{TM} i3 $-$ 540 CPU 3.06 GHz.

Table 3.2: A performance comparison of the geometrical motion planners

Domain	Execution time in ms		Path length in m		Number of way points		Success probability	
	RRT	A*	RRT	A*	RRT	A*	RRT	A*
Scenario 1	2.7	1294.2	3.8949	3.8949	2.0	2.0	100 %	100 %
Scenario 2	4.5	1038.5	4.6507	4.0752	3.0740	3.0	100 %	100 %
Scenario 3	17.7	756.2	4.1680	2.7726	3.9420	4.0	100 %	100 %
Scenario 4	84.7	927.3	6.4849	5.2668	6.0470	6.0	100 %	100 %
Scenario 5	842.5	2066.8	10.7481	9.5145	10.4887	10.0	93.3%	100 %
Scenario 6	42.6	1555.8	6.1165	4.3886	6.3150	8.0	100 %	100 %
Scenario 7	32.1	919.1	5.1078	3.6823	5.5050	5.0	100 %	100 %
Scenario 8	59.4	1288.1	5.8989	4.1010	6.7460	9.0	100 %	100 %
Scenario 9	33.5	1330.7	3.8599	2.9374	4.4210	5.0	100 %	100 %

The performances of both geometrical planners on the benchmark testing environments are summarized in Table 3.2 and visualized in Figure 3.7. Apparently, the RRT geometrical planner has a considerable advantage over the A* grid-based geometrical planner in execution time. Particularly, the RRT geometrical planner needs less than 100 ms in most scenarios except the Zigzag scenario 5. Due to space connectivity in the Zigzag scenario 5, the RRT geometrical planner requires the largest execution time of 842.5 ms. This effect has been also reported [11]. Moreover, the lowest execution times of the RRT geometrical planner are 2.7ms and 4.5 ms for Scenarios 1 and 2, while the lowest execution times for the A* grid-based geometrical planner are 756.2 ms and 919.1 ms for Scenarios 3 and 7. In general, the execution time for both geometrical planners are not proportional to the numbers of obstacles. For the RRT geometrical planner, the execution times depend on the space connectivity and expansiveness [60]. In contrast to the RRT geometrical planner, the execution time of the A^* grid-based geometrical planner depends on the chosen heuristic function and the number of the explored cells.

Since the A* grid-based geometrical planner uses the Euclidean distance as the cost function, the path lengths of the A* grid-based geometrical planner are obviously shorter than those of the RRT geometrical planner in all scenarios. Nevertheless, the A* grid-based geometrical planner does not guarantee a minimal number of way points as shown in Table 3.2, which has a profound negative influence on the trajectory optimization algorithm introduced in Chapter 4. In term of success probability, the geometrical planners do not fail to find a solution in most cases except Scenario 5. Particularly, the success rate of the RRT geometrical planner is only 93.3% in Scenario 5 due to the space connectivity. However, the success probability will increase to 98.1% if the maximal allowable iteration number is 2000.

3.5 Conclusion

In this chapter, the RRT and A^* grid-based geometrical planners are introduced. Furthermore, the working principles of these geometrical planners are fully explained and all ingredient algorithms are described in detail. Lastly, the performances of both geometrical planners are compared across several environment scenarios. In conclusion, the RRT geometrical planner is favorable against the A^* grid-based geometrical planner. Indeed, the RRT geometrical planner requires lower execution times in all scenarios than the A^* grid-based geometrical planner. Nevertheless, the RRT planner delivers stochastic results, which are not suitable for some applications.

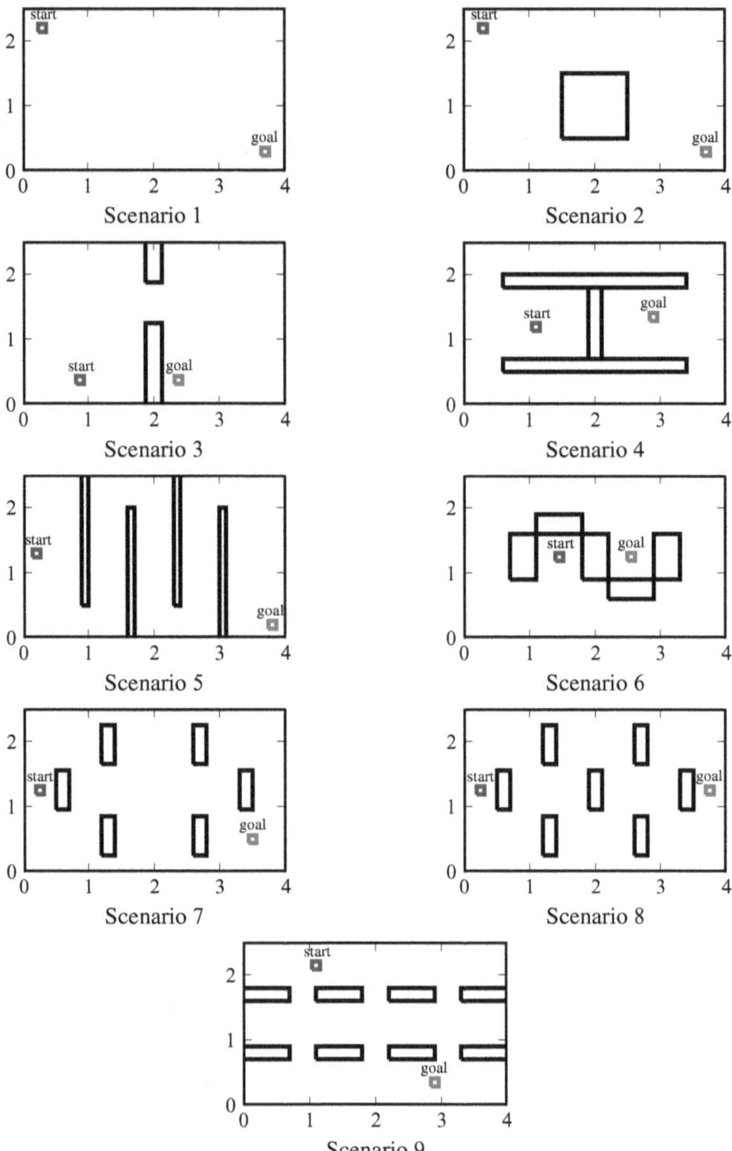

Figure 3.6: All benchmark testing environment domains, which covers a broad range of diversity and space connectivity. Each rectangular scenario has a width of $2.5\ m$ and a length of $4.0\ m$ Additionally, robots are modeled by a square, which has a width of $0.12\ m$. The start positions are marked by blue rectangles, while the goal positions are marked by red rectangles.

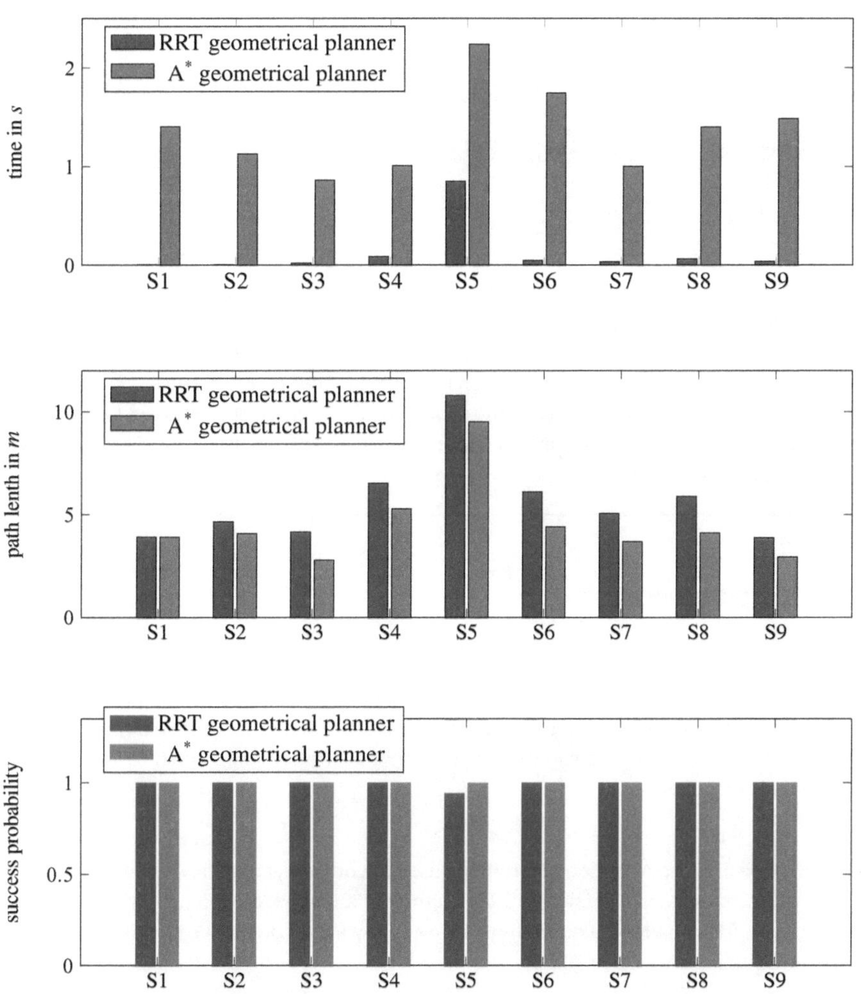

Figure 3.7: A performance comparison of the A* grid-based and RRT geometrical planners across several environment domains. S1, S2, ..., S9 denote the scenario enumeration.

4 A High-order Near Time-optimal Trajectory Generation Algorithm

This chapter presents a novel high-order near time-optimal trajectory generation algorithm for two-dimensional robots. The algorithm is able to generate trajectories for point-to-point movements, which can be embedded in factory and logistic automation. Moreover, the generated trajectories are not only near time-optimal but also bounded by high-order time derivatives. Furthermore, discretization effects as well as nonlinear constraints due to stationary obstacles are incorporated into the algorithm. Finally, low computational burden and industrial applicability of the algorithm are demonstrated by numerical examples.

4.1 Introduction

Two-dimensional robots are widely applied in various factory automation applications. For instance, they are deployed in positioning systems such as CNC machines, assembly facilities, 3-D printing and the material handling systems mentioned in Chapter 1. In general, performance of any two-dimensional robots is specified by a motion control system, which consists of the two fundamental components:

- *Feedback controller:* A motion controller is used to compensate unknown disturbances as well as unmodelled dynamics. Classically, PID controllers are applied, which require low computational burden [54]. However, a recent trend is to employ model predictive controllers (MPC) with high computational complexity [22].

- *Trajectory planning:* Feasible trajectories are computed, which not only satisfy kinematic and safety constraints but also are optimal with respect to relevant performance objectives. Consequently, the trajectory generation algorithm has often high computational burden.

As aforementioned, a near time-optimal trajectory generation algorithm is developed in this chapter. In the following, the constraints as well as the requirements of the trajectory optimization problem are recapitulated.

• *Boundary Conditions:* In this chapter, third-order trajectories are considered. However, the presented algorithm can be even extended to generate higher-order trajectories. The initial state with non-zero initial velocity and acceleration is given as

$$\mathbf{x}_0 = \begin{bmatrix} q_{x,\text{start}} & \dot{q}_{x,0} & \ddot{q}_{x,0} & q_{y,\text{start}} & \dot{q}_{y,0} & \ddot{q}_{y,0} \end{bmatrix}^T .$$

The end state is specified by zero velocity and acceleration as

$$\mathbf{x}_e = \begin{bmatrix} q_{x,\text{goal}} & 0 & 0 & q_{y,\text{goal}} & 0 & 0 \end{bmatrix}^T .$$

• *Kinematic Constraints:* The velocity, acceleration and jerk for both x- and y- dimensions are bounded by the vectors $\mathbf{v}_{\max}$, $\mathbf{a}_{\max}$ and $\mathbf{j}_{\max}$.

• *Obstacles:* A list of obstacles is given as $\mathbf{O} = \{O_1, O_2, \ldots, O_{N_o}\}$, which are defined by a set of convex polygons as mentioned in Chapter 1.

• *Time Discretization Effect:* The time discretization effect must be considered for high-performance motion control systems [54]. In fact, the third time derivative $\dddot{q}_\xi(t)$ of the coordinate ξ, $\xi \in \{x, y\}$, is only discontinuous at time stamps t_d, which are integers of the sampling time T_s. In other words, one has that $t_\mathrm{d} = n_\mathrm{d} T_\mathrm{s}$ and $n_\mathrm{d} \in \mathbb{N}$, $\forall t_\mathrm{d} \in \mathcal{F}(\dddot{q}_\xi(t))$, where $\mathcal{F}(\dddot{q}_\xi(t))$ is the set of the discontinuous time stamps.

• *Complexity:* Since the desired end position can be adjusted and the working environment can be changed, trajectories are replanned on the fly. Therefore, the algorithm should have a low computational complexity.

• *Performance:* In order to maximize productivity, the total travel time is minimized.

In conclusion, the trajectory optimization problem (1.1) for two-dimensional robots can be reformulated and expressed as

$$\min_{\mathbf{q}(t),N} \quad N \tag{4.1a}$$

$$\text{subject to} \quad N \in \mathbb{N}, \qquad T = NT_\mathrm{s}, \quad \mathbf{q}(t) = \begin{bmatrix} q_x(t) \\ q_y(t) \end{bmatrix}, \tag{4.1b}$$

$$\mathbf{q}(0) = \mathbf{q}_{\text{start}}, \qquad \mathbf{q}(T) = \mathbf{q}_{\text{goal}}, \tag{4.1c}$$

$$\dot{\mathbf{q}}(0) = \begin{bmatrix} v_{0,x} \\ v_{0,y} \end{bmatrix}, \quad \dot{\mathbf{q}}(T) = \begin{bmatrix} 0 \\ 0 \end{bmatrix}, \tag{4.1d}$$

$$\ddot{\mathbf{q}}(0) = \begin{bmatrix} a_{0,x} \\ a_{0,y} \end{bmatrix}, \quad \ddot{\mathbf{q}}(T) = \begin{bmatrix} 0 \\ 0 \end{bmatrix}, \tag{4.1e}$$

$$- \mathbf{v}_{max} \leq \dot{\mathbf{q}}(t) \leq \mathbf{v}_{max}, \qquad (4.1f)$$

$$- \mathbf{a}_{max} \leq \ddot{\mathbf{q}}(t) \leq \mathbf{a}_{max}, \qquad (4.1g)$$

$$- \mathbf{j}_{max} \leq \dddot{\mathbf{q}}(t) \leq \mathbf{j}_{max}, \qquad (4.1h)$$

$$O(\mathbf{q}(t)) \cap O_i = \varnothing, \quad \forall i \in \{1, \ldots, N_o\}, \qquad (4.1i)$$

$$t_d = n_d T_s, \quad n_d \in \mathbb{N}, \quad \forall t_d \in \mathcal{F}(\dddot{q}_\xi(t)), \qquad (4.1j)$$

$$\forall t \in [0 \quad T], \qquad (4.1k)$$

where T_s is the sampling time.

In the following, a near time-optimal trajectory generation algorithm for two-dimensional robots is presented, which is able to handle the drawbacks of all approaches mentioned in the literature review. The chapter is organized by the subsequent sections:

- **Trajectory Optimization Based on the Decompositional Approach:** The decompositional approach and fundamental ingredients of the high-order near time-optimal trajectory generation algorithm is described.

- **Smooth Path Generation Using Quintic Bézier Splines:** The smooth path generation based on parameterization of quintic Bézier splines is presented.

- **A Near Time-optimal Trajectory Fitting Algorithm:** An efficient algorithm is proposed, which is capable of mitigating infinite jerks and generating a near time-optimal trajectory along a given smooth path. Furthermore, the time discretization effect is also considered by this algorithm. Additionally, a termination analysis of the algorithm is provided.

- **Gradient-free Trajectory Optimization:** In order to improve the performance, path parameters are optimized by using a gradient-free algorithm introduced in this section.

- **Illustrative Examples:** The efficiency and applicability of the trajectory generation algorithm are demonstrated by numerical examples.

- **Conclusion:** Finally, the chapter is concluded with a discussion about the presented results.

4.2 Trajectory Optimization Based on the Decompositional Approach

As shown in Figure 4.1, the approach decomposes the time-optimal trajectory generation into multiple subproblems, which are solved by distinct algorithms. Obviously, the choice of these algorithms will have influence on performance and characteristics of the approach. In fact, the near time-optimal trajectory generation algorithm is comprised of a geometrical path planner and a trajectory optimization algorithm. Moreover, the trajectory optimization algorithm is decoupled into three elements, which are a path generation algorithm using quintic Bézier splines, a near time-optimal trajectory fitting algorithm as well as a gradient-free optimization algorithm. In the following, these ingredient algorithms and their complementary roles are explained.

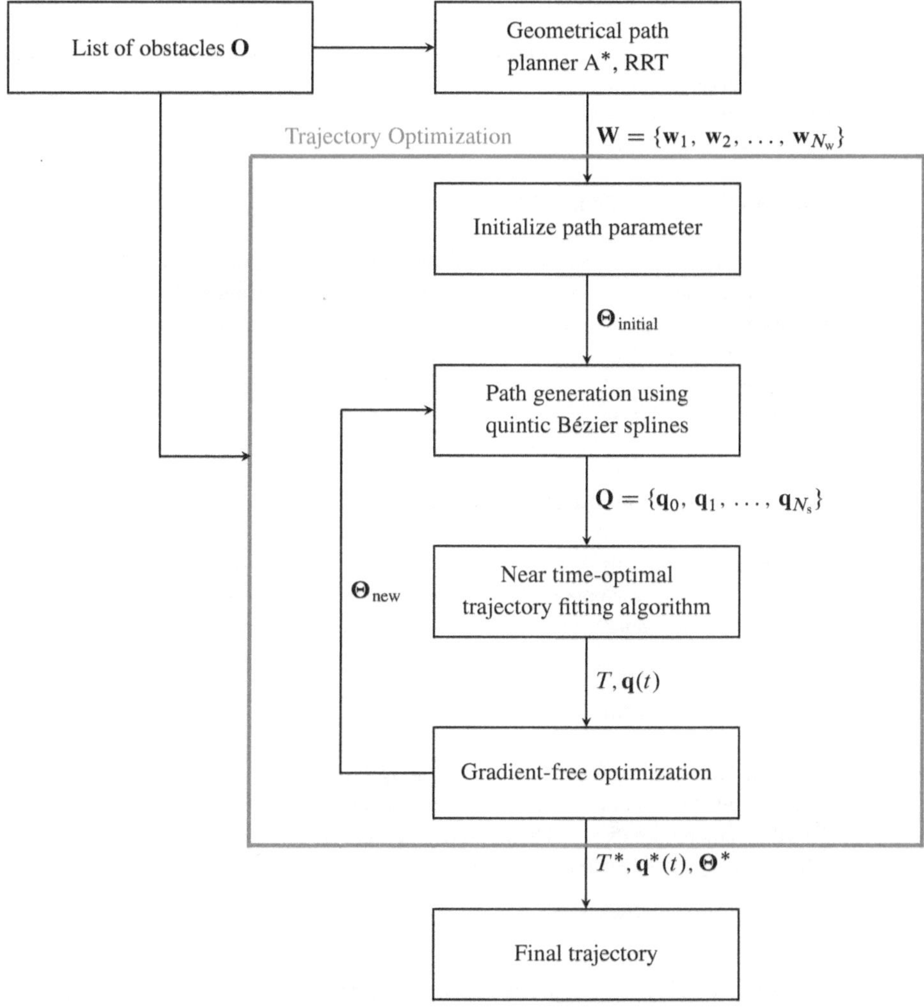

Figure 4.1: Trajectory Optimization

4.2.1 Geometrical Path Planner

Given a list of obstacles $\mathbf{O}$, a geometrical planner is employed to find a list of way points $\mathbf{W} = \{\mathbf{w}_1, \mathbf{w}_2, \ldots, \mathbf{w}_{N_w}\}$ from a start to a goal with $\mathbf{w}_1 = \mathbf{q}_{\text{start}}$ and $\mathbf{w}_{N_w} = \mathbf{q}_{\text{goal}}$, where each straight line connecting two consecutive way points $\mathbf{w}_i$ and $\mathbf{w}_{i+1}$, $\forall i = \{1, \ldots, N_w - 1\}$, is geometrically collision-free. Theoretically, no panacea geometrical planner exists. In other words, geometrical planners are either resolution (A^*) or probabilistic (RRT) complete. Consequently, the completeness of our approach depends solely on the chosen geometrical planner. Moreover, both A^* grid-based and RRT geometrical planners are very efficient for finding a geometrical path in two-dimensional configuration spaces as aforementioned in Chapter 3. However, the geometrical path planner is not the main scope of this chapter.

4.2.2   Path Generation Using Quintic Bézier Splines

The geometrical planner delivers non-smooth unnecessary motions. Furthermore, the kinematic constraints $\mathbf{v}_{max}$, $\mathbf{a}_{max}$ and $\mathbf{j}_{max}$ are not considered in the geometrical planners. Therefore, trajectories are smoothed and optimized in the second step. Based on the list of way points $\mathbf{W}$, a smooth path $\mathcal{S}$ is parametrized by using quintic Bézier splines. In other words, the smooth path $\mathcal{S}$ is defined by a parameter $\mathbf{\Theta} \in \mathbb{R}^{3 \times (N_w - 2)}$, where N_w is the number of way points. Eventually, an ultimate trajectory is obtained by optimizing the parameter $\mathbf{\Theta}$.

4.2.3   Near Time-optimal Trajectory Fitting Algorithm

The smooth path $\mathcal{S}$ is sampled by $N_s + 1$ configurations. Afterwards, a list of sampling configurations $\mathbf{Q} = \{\mathbf{q}_0, \mathbf{q}_1, \ldots, \mathbf{q}_{N_s}\}$ is obtained. By fitting cubic splines, a near time-optimal trajectory along the smooth path $\mathcal{S}$ is generated, which satisfies all high-order time derivative constraints.

4.2.4   Gradient-Free Optimization

Finally, a gradient-free optimization algorithm is deployed in order to minimize trajectory travel time T. In the gradient-free optimization algorithm, the search direction is found and adaptively changed based on computing the trajectory and its cost for an alternative parameter $\mathbf{\Theta}_{new}$ recursively as shown in Figure 4.1. If planning time is up, an ultimate trajectory $\mathbf{q}^*(t)$ as well as its parameter $\mathbf{\Theta}^*$ are delivered.

4.3   Smooth Path Generation Using Quintic Bézier Splines

Smooth path generation using parametric splines is prevalent in robotic motion planning and can be found in extensive researches [67, 92, 93, 111]. Several spline types are applied in trajectory optimization [92]. In this section, we will explain why the quintic Bézier splines are chosen. Afterwards, spline parametrization based on the way point list $\mathbf{W}$ is described. Furthermore, the dimension of the optimized path parameter $\mathbf{\Theta}$ can be reduced by using similar heuristics as in the

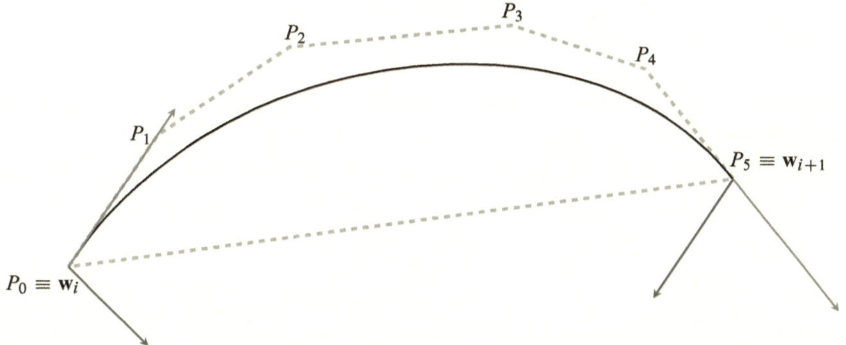

Figure 4.2: Quintic Bézier spline segment

related work [92].

The quintic Bézier splines are suitable for solving the time-optimal trajectory planning problem (4.1) due to their following important properties. By using Bernstein polynomials, a quintic Bézier spline is defined as

$$\mathbf{q}(u) = (1-u)^5 P_0 + 5(1-u)^4 u P_1 + 10(1-u)^3 u^2 P_2$$
$$+ 10(1-u)^2 u^3 P_3 + 5(1-u)u^4 P_4 + u^5 P_5,$$

where $u \in [0 \quad 1]$ is the spline internal parameter and $P_i \in \mathbb{R}^2$, $i \in \{0, 1, \ldots, 5\}$, are the control vertices. First, the start and end points of a quintic Bézier spline are constituted by the first and last control vertices as

$$\mathbf{q}(u = 0) = P_0 = \mathbf{w}_i, \quad \mathbf{q}(u = 1) = P_5 = \mathbf{w}_{i+1},$$

which are respectively the way points $\mathbf{w}_i$ and $\mathbf{w}_{i+1}$ given by the geometrical planners. Furthermore, the first and second derivatives at the start and end points are induced by the chosen control vertices as

$$\frac{d\mathbf{q}}{du}(u = 0) = \mathcal{S}_i'(0) = 5(P_1 - P_0), \tag{4.2}$$

$$\frac{d^2\mathbf{q}}{du^2}(u = 0) = \mathcal{S}_i''(0) = 20(P_0 - 2P_1 + P_2), \tag{4.3}$$

$$\frac{d\mathbf{q}}{du}(u = 1) = \mathcal{S}_i'(1) = 5(P_5 - P_4), \tag{4.4}$$

$$\frac{d^2\mathbf{q}}{du^2}(u = 1) = \mathcal{S}_i''(1) = 20(P_3 - 2P_4 + P_5), \tag{4.5}$$

which is a paramount property for specifying the initial and final states $\dot{\mathbf{q}}(0)$, $\dot{\mathbf{q}}(T)$, $\ddot{\mathbf{q}}(0)$, $\ddot{\mathbf{q}}(T)$ in the optimal control problem. As shown in Figure 4.2, a Bézier Spline has a convex hull property, which implies that a spline path always lies within a convex hull defined by the control vertices P_i. Consequently, one is able to deploy the separating hyperplane Theorem 3.1 in order to impose obstacle avoidance constraints. Ultimately, each configuration connection $\mathbf{w}_i\mathbf{w}_{i+1}$, $\forall i \in \{1, \ldots, N_\mathbf{w} - 1\}$, can be smoothed by a quintic Bézier spline. From (4.2), (4.3), (4.4) and (4.5), a smooth path will be determined if the first and second derivative vectors at each way point $\mathbf{w}_i$ are specified.

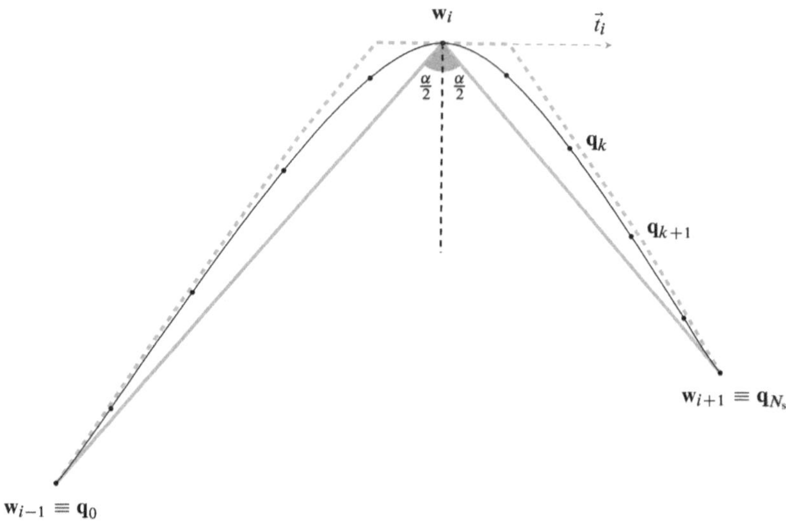

Figure 4.3: Smooth quintic Bézier splines path generation and parametrization by using heuristics

Moreover, the heuristics of choosing the first and second derivative vectors for smooth path initialization and optimization based on the quintic Bézier splines have been extensively studied. For more details, interested readers are referred to the related works [39, 91, 92, 93]. In the following, these heuristics are briefly mentioned. Subsequently, a smooth path parametrization from a list of way points is introduced. First of all, the tangent direction at an inner way point $\mathbf{w}_i$ is set to be orthogonal to the bisector of the angle $\alpha = \angle\mathbf{w}_{i-1}\mathbf{w}_i\mathbf{w}_{i+1}$ inscribed by the previous and next way points. Moreover, the amplitude of the tangent is initialized by a half of minimal Euclidean distances to neighbor way points as

$$|\vec{t}_{\mathbf{w}_i}| = \frac{1}{2} \min\{|\mathbf{w}_{i-1}\mathbf{w}_i|, |\mathbf{w}_i\mathbf{w}_{i+1}|\}.$$

For path initialization, these tangent amplitudes can be iteratively reduced or even set to be zero in order to obtain a feasible collision-free path. On one hand, the larger the tangent amplitudes are, the smoother path can be generated. On the other hand, a large tangent amplitude can lead to a collision path. Consequently, these amplitudes are set as optimization variables. Moreover, another objective is to minimize the integral of the second derivative along the path $\mathcal{S}$ in order to yield an optimal smooth trajectory, since it is a crude measure of a total curvature along a path. Nevertheless, minimizing the curvature integral is a property of cubic splines [39, 91, 92, 93]. Therefore, the second derivative vector is chosen by mimicking behaviors of cubic splines. Moreover, cubic Bézier splines are determined given the way points as well as their tangents. However, applying the cubic Bézier splines induces discontinuities of the curvatures at the given way points. Consequently, the curvature at an inner way point $\mathbf{w}_i$ is retrieved by weighting the curvature of the cubic Bézier splines in the following.

Given three way points and tangents $\{\vec{\mathbf{w}}_{i-1}, \vec{t}_{\mathbf{w}i-1}\}$, $\{\vec{\mathbf{w}}_i, \vec{t}_{\mathbf{w}_i}\}$ and $\{\vec{\mathbf{w}}_{i+1}, \vec{t}_{\mathbf{w}_{i+1}}\}$, the second derivatives at the way point $\mathbf{w}_i$ are

$$\mathcal{S}''_{\mathbf{w}_{i-1}\mathbf{w}_i}(u = 1) = \mathcal{S}''_{i-1}(1) = 6\vec{\mathbf{w}}_{i-1} + 2\vec{t}_{i-1} + 4\vec{t}_i - 6\vec{\mathbf{w}}_i, \tag{4.6}$$

$$\mathcal{S}''_{\mathbf{w}_i\mathbf{w}_{i+1}}(u = 0) = \mathcal{S}''_i(0) = -6\vec{\mathbf{w}}_i - 4\vec{t}_{i-1} - 2\vec{t}_{i+1} + 6\vec{\mathbf{w}}_{i+1}, \tag{4.7}$$

obtained by applying the cubic Bézier splines. The second derivative vector at an inner point $\mathbf{w}_i$ is an average of the above two second derivatives, which is weighted inversely proportional with the Euclidean distances to the neighbor way points as

$$\vec{a}_{\mathbf{w}_i} = \frac{|\mathbf{w}_i\mathbf{w}_{i+1}|}{|\mathbf{w}_{i-1}\mathbf{w}_i| + |\mathbf{w}_i\mathbf{w}_{i+1}|}\mathcal{S}''_{i-1}(1) + \frac{|\mathbf{w}_{i-1}\mathbf{w}_i|}{|\mathbf{w}_{i-1}\mathbf{w}_i| + |\mathbf{w}_i\mathbf{w}_{i+1}|}\mathcal{S}''_i(0).$$

For the first and last way points, the tangent directions are either given by the initial of the optimization problem (4.1) or set to be the straight line to the next or from previous way points. Furthermore, the tangent amplitudes are heuristically set to be half of the the Euclidean distance to the next way point or from the previous way point as $|\vec{t}_{\mathbf{w}_1}| = \frac{1}{2}|\mathbf{w}_1\mathbf{w}_2|$, $|\vec{t}_{\mathbf{w}_{N_w}}| = \frac{1}{2}|\mathbf{w}_{N_w-1}\mathbf{w}_{N_w}|$. Similarly, the second derivative directions of the first and last way points are either predetermined by the initial condition or correspondingly equal to the values defined by the cubic Bézier splines.

By employing the aforementioned heuristics, the number of the optimization variables can be drastically reduced. In fact, three variables for each inner way point are optimized, which are respectively the tangent amplitude $t_{\mathbf{w}_i}$ and the way point coordinates. In other words, the number of the optimization path parameters Θ is $3(N_w - 2)$ for a list of N_w way points. In the next section, an efficient algorithm fitting a near time-optimal trajectory along the smooth path $\mathcal{S}$ is introduced.

4.4 A Near Time-optimal Trajectory Fitting Algorithm

Generally speaking, the best trajectory is obtained if both the smooth path $\mathcal{S}$ and the trajectory along the smooth path are optimal. In this section, the time-optimal trajectory generation along a specified path subject to kinematic constraints is considered, which is a fundamental problem and encountered in many applications. Consequently, several solutions to the problem are proposed [73, 88]. Basically, these solutions are based on the bang-bang nature or the Pontryagin's maximum principle of the time-optimal velocity profile. Particularly, switching curves are found by solving switching points on phase-planes analytically and applying numerical forward and backward integration methods [7, 73, 88]. On the other hand, the problem can be interpreted and solved by versatile robust second-order cone programming [104]. Nonetheless, these approaches have two main drawbacks. Firstly, the time discretization effect is not taken into account. Secondly, the jerk constraint $j_{\max}$ is not imposed, which nonetheless leads to a non-convex optimization problem [104]. These drawbacks are mitigated by an efficient simple algorithm proposed in the following.

By sampling the smooth path $\mathcal{S}$, one obtains a list of points $\mathbf{Q} = \{\mathbf{q}_0, \mathbf{q}_1, \ldots, \mathbf{q}_{N_s}\}$ as shown in Figure 4.3, where $\mathbf{q}_0$ and $\mathbf{q}_{N_s}$ are respectively the initial and final positions. Since the jerk constraint is considered, a reasonable idea is to find a cubic spline interpolation through these points

$$\{(t_0 = 0, \mathbf{q}_0), (t_1, \mathbf{q}_1), \ldots, (t_k, \mathbf{q}_k), \ldots, (T, \mathbf{q}_{N_s})\},$$

where t_k is respectively the time parameter. If

$$M_{k,\xi} = \ddot{q}_\xi(t_k), \quad \forall\, k \in \{0, 1, \ldots, N_s\}, \quad \forall\, \xi \in \{x, y\},$$

is denoted as the acceleration at the samples $\mathbf{q}_k$ of the coordinate ξ, then a cubic spline trajectory of the coordinate ξ in the interval $[\mathbf{q}_{k-1} \quad \mathbf{q}_k]$ can be written as

$$
\begin{aligned}
q_\xi(t) = S_{k,\xi}(t) = {}& M_{k-1,\xi}\frac{(t_k - t)^3}{6h_k} + \left(q_{k-1,\xi} - M_{k-1,\xi}\frac{h_k^2}{6}\right)\frac{t_k - t}{h_k} \\
& + M_{k,\xi}\frac{(t - t_{k-1})^3}{6h_k} + \left(q_{k,\xi} - M_{k,\xi}\frac{h_k^2}{6}\right)\frac{t - t_{k-1}}{h_k},
\end{aligned}
\tag{4.8}
$$

where $t_{k-1} \leq t \leq t_k$ and $h_k = t_k - t_{k-1}$ is the time interval. By fitting the cubic splines through these $N_s + 1$ sampling points, $4N_s$ coefficients are available while $4N_s + 2$ conditions including the initial and final velocities as well as accelerations must be satisfied. Therefore, the first and last segments are approximated by the following quintic splines

$$
q_\xi(t) = S_{1,\xi}(t) := q_\xi(t) - (M_{0,\xi} - a_{0,\xi})\frac{t^2(t_1 - t)^3}{2h_1^3}, \quad \forall\, 0 \leq t \leq t_1,
\tag{4.9}
$$

$$
q_\xi(t) = S_{N_s,\xi}(t) := q_\xi(t) - M_{N_s,\xi}\frac{(t - t_{N_s-1})^3(T - t)^2}{2h_{N_s}^3}, \quad \forall\, t_{N_s-1} \leq t \leq T,
\tag{4.10}
$$

where $a_{0,\xi}$ is the initial acceleration, T is the total travel time. While the acceleration continuity is inherently ensured by (4.8), the velocity continuity is guaranteed by the additional equation

$$
\begin{aligned}
& \dot{S}_{k,\xi}(t = t_k) = \dot{S}_{k+1,\xi}(t = t_k) \\
\Leftrightarrow\; & \frac{1}{3}M_{k,\xi}h_k + \frac{1}{6}M_{k-1,\xi}h_k + \frac{q_{k,\xi} - q_{k-1,\xi}}{h_k} = -\frac{1}{3}M_{k,\xi}h_{k+1} - \frac{1}{6}M_{k+1,\xi}h_{k+1} + \frac{q_{k+1,\xi} - q_{k,\xi}}{h_{k+1}} \\
\Leftrightarrow\; & \frac{h_k}{h_k + h_{k+1}}M_{k-1,\xi} + 2M_{k,\xi} + \frac{h_{k+1}}{h_k + h_{k+1}}M_{k+1,\xi} = \underbrace{\frac{6}{h_k + h_{k+1}}\left(\frac{q_{k+1,\xi} - q_{k,\xi}}{h_{k+1}} - \frac{q_{k,\xi} - q_{k-1,\xi}}{h_k}\right)}_{d_{k,\xi}}.
\end{aligned}
\tag{4.11}
$$

Additionally, one has from the initial and final velocity conditions (4.1d) that

$$2M_{0,\xi} + M_{1,\xi} = \frac{6(q_{1,\xi} - q_{0,\xi})}{h_1^2} - \frac{6v_0}{h_1} = d_{0,\xi},$$

$$M_{N_s-1,\xi} + 2M_{N_s,\xi} = -\frac{6(q_{N_s,\xi} - q_{N_s-1,\xi})}{h_{N_s}^2} = d_{N_s,\xi}. \tag{4.12}$$

From (4.11) and (4.12), one obtains

$$\underbrace{\begin{bmatrix} 2 & 1 & 0 & \cdots & 0 & 0 & 0 \\ \mu_1 & 2 & 1-\mu_1 & \cdots & 0 & 0 & 0 \\ 0 & \mu_2 & 2 & 1-\mu_2 & \cdots & 0 & 0 \\ \vdots & \vdots & \vdots & \vdots & \vdots & \vdots & \vdots \\ 0 & 0 & 0 & \cdots & \mu_{N_s-1} & 2 & 1-\mu_{N_s-1} \\ 0 & 0 & 0 & \cdots & 0 & 1 & 2 \end{bmatrix}}_{\mathbf{A}} \underbrace{\begin{bmatrix} M_{0,\xi} \\ M_{1,\xi} \\ M_{2,\xi} \\ \vdots \\ M_{N_s-1,\xi} \\ M_{N_s,\xi} \end{bmatrix}}_{\mathbf{M}_\xi} = \underbrace{\begin{bmatrix} d_{0,\xi} \\ d_{1,\xi} \\ d_{2,\xi} \\ \vdots \\ d_{N_s-1,\xi} \\ d_{N_s,\xi} \end{bmatrix}}_{\mathbf{d}_\xi}, \tag{4.13}$$

where $0 < \mu_k = \frac{h_k}{h_k + h_{k+1}} < 1$. By applying cubic spline fitting, the minimum-time trajectory generation along a specified path can be relaxed and reformulated as

$$\min_{h_k, M_k} \quad T = \sum_{k=1}^{N_s} h_k \tag{4.14}$$

$$\text{subject to} \quad (4.8), (4.9), (4.10), (4.13), (4.1f), (4.1g), (4.1h), \quad \forall \, \xi \in \{x, y\},$$

$$h_k = n_k T_s, \quad n_k \in \mathbb{N}, \quad n_k \geq 1, \quad \forall \, k \in \{1, 2, \ldots, N_s\}.$$

Furthermore, the time discretization effect is considered by incorporating the condition $h_k = n_k T_s$, $n_k \in \mathbb{N}$, into the optimization problem (4.14). In other words, the jerk $\dddot{q}_\xi(t)$ is only discontinuous at the time stamp t_k, which is an integer of the sampling time T_s. However, solving the optimization problem (4.14) is cumbersome. An elaborate search algorithm is introduced by Chun Lin, which obviously has high computational burden [61]. Therefore, an efficient Algorithm 4.1 is proposed, which works well in practice [39, 109]. Since minimum-time trajectories are desired, the time intervals h_k are initialized such that maximal allowable velocity along the path is encouraged. In fact, the initial time intervals are determined by

$$h_1 = \max_{\xi \in \{x,y\}} \lceil (\frac{3}{2} - \frac{v_{0,\xi}\,\text{sign}(q_{1,\xi} - q_{0,\xi})}{2v_{\max}}) \frac{|q_{1,\xi} - q_{0,\xi}|}{v_{\max} T_s} \rceil T_s,$$

$$h_k = \max_{\xi \in \{x,y\}} \lceil \frac{|q_{k,\xi} - q_{k-1,\xi}|}{v_{\max} T_s} \rceil T_s, \qquad \forall \, k \in \{2, \ldots, N_s - 1\},$$

$$h_{N_s} = \max_{\xi \in \{x,y\}} \lceil \frac{3|q_{N_s,\xi} - q_{N_s-1,\xi}|}{2v_{\max} T_s} \rceil T_s,$$

where $\lceil \epsilon \rceil$ denotes the least integer greater than ϵ. As long as the kinematic constraints (4.1f), (4.1g), (4.1h) are not fulfilled on the segment $S_{k,\xi}$, the time interval h_k on the same segment $S_{k,\xi}$ is increased by T_s as shown in lines 8-10 of Algorithm 4.1. Furthermore, the maximal velocity

$\dot{S}_{k,\xi}(t)$, acceleration $\ddot{S}_{k,\xi}(t)$ and jerk $\dddot{S}_{k,\xi}(t)$ can be analytically computed in each time interval $[t_{k-1} \quad t_{k-1} + h_k]$, since each spline segment $S_{k,\xi}(t)$ is a either cubic or quintic polynomial of the time parameter t. Moreover, the spline coefficient vector can be obtained from (4.13) as

$$\mathbf{M}_\xi = \mathbf{A}^{-1}\mathbf{d}_\xi, \quad \forall\, \xi \in \{x, y\}.$$

Algorithm 4.1 Determining the time intervals $\mathbf{h} = \{h_1, h_2, \ldots, h_{N_s}\}$

    Input: A list of of sample points : $\mathbf{Q} = \{\mathbf{q}_0, \mathbf{q}_1, \ldots, \mathbf{q}_{N_s}\}$, the kinematic constraints $\mathbf{v}_{\max}, \mathbf{a}_{\max}, \mathbf{j}_{\max}$
    Output: The time interval $\mathbf{h}$ and the spline coefficients $\{M_{0,\xi}, M_{1,\xi}, \ldots, M_{N_s,\xi}\}$

1: **Initialization:**

2:    $h_1 = \max\limits_{\xi \in \{x,y\}} \lceil (\frac{3}{2} - \frac{v_0 \,\mathrm{sign}(q_{1,\xi} - q_{0,\xi})}{2v_{\max,\xi}}) \frac{|q_{1,\xi} - q_{0,\xi}|}{v_{\max,\xi} T_s} \rceil T_s;$

3:    $h_k = \max\limits_{\xi \in \{x,y\}} \lceil \frac{|q_{k,\xi} - q_{k-1,\xi}|}{v_{\max,\xi} T_s} \rceil T_s, \quad \forall\, k = 2, \ldots, N_s - 1;$

4:    $h_{N_s} = \max\limits_{\xi \in \{x,y\}} \lceil \frac{3|q_{N_s,\xi} - q_{N_s-1,\xi}|}{2v_{\max,\xi} T_s} \rceil T_s;$

5:    $\mathbf{M}_\xi = \mathbf{A}^{-1}\mathbf{d}_\xi, \quad \forall\, \xi \in \{x, y\};$

6: **repeat**

7:    **for all** $k = 1 : 1 : N_s$ **do**

8:       **if** $\big((\max|\dot{S}_{k,\xi}(t)| > v_{\max,\xi}) \vee (\max|\ddot{S}_{k,\xi}(t)| > a_{\max,\xi}) \vee (\max|\dddot{S}_{k,\xi}(t)| > j_{\max,\xi})\big)$ **then**

9:          $h_k \quad h_k + T_s;$

10:       **end if**

11:    **end for**

12:    $\mathbf{M}_\xi = \mathbf{A}^{-1}\mathbf{d}_\xi, \quad \forall\, \xi \in \{x, y\};$

13: **until** $(\max|\dot{q}_\xi(t)| \leq v_{\max,\xi}) \wedge (\max|\ddot{q}_\xi(t)| \leq a_{\max,\xi}) \wedge (\max|\dddot{q}_\xi(t)| \leq j_{\max,\xi})$

Furthermore, the matrix $\mathbf{A}$ is strictly diagonal dominant and consequently invertible. Additionally, computing the inverse matrix $\mathbf{A}^{-1}$ has a complexity of $\mathcal{O}(N_s^3)$ operations. Nevertheless, Thomas' back-substitution algorithm is employed to obtain the solution $\mathbf{M}_\xi$ in $\mathcal{O}(N_s)$ operations by utilizing the tridiagonal band structure of the matrix $\mathbf{A}$ [19]. By using the Thomas' back-substitution algorithm, the solution $\mathbf{M}_\xi$ is obtained by using the following recursive first-order formulas

$$a_{k,\xi} := -\frac{1 - \mu_k}{2 + \mu_k a_{k-1,\xi}}, \; a_{-1,\xi} = 0, \; \mu_0 = 0, \quad \forall\, k = \{0, 1, \ldots, N_s\},$$

$$b_{k,\xi} := \frac{d_{k,\xi} - \mu_k b_{k-1,\xi}}{2 + \mu_k a_{k-1,\xi}}, \quad b_{-1,\xi} = 0, \; \mu_{N_s} = 1, \quad \forall\, k = \{0, 1, \ldots, N_s\},$$

$$M_{k,\xi} := a_{k,\xi} M_{k+1,\xi} + b_{k,\xi}, \quad M_{N_s,\xi} = b_{N_s,\xi}, \quad \forall\, k = \{N_s - 1, \ldots, 0\}.$$

Moreover, Algorithm 4.1 terminates as the number of iterations tends to infinity. In other words, the iteration number of the repeat loops in Algorithm 4.1 is limited, if some relaxed additional assumptions are fulfilled. For the sake of simplicity, the following assumptions are made.

Assumption 4.1 *The quintic spline approximation of the first and last segments can be neglected for the termination analysis of Algorithm 4.1.*

Assumption 4.2 *After a finite number of the repeat loops, the following conditions*

$$\frac{2|q_{k,\xi} - q_{k-1,\xi}|}{3h_k} \leq v_{max}, \quad \forall\, k = \{2, \ldots, N_s\},$$

$$\frac{2}{3}\left|\frac{q_{1,\xi} - q_{0,\xi}}{h_1} + \frac{1}{12}d_{0,\xi}h_1\right| \leq v_{max}, \quad \forall\, \xi \in \{x, y\}, \tag{4.15}$$

are satisfied.

Remark 4.1 *For $k = N_s$, the condition (4.15) holds at the initialization of Algorithm 4.1. Consequently, Assumption 4.2 is satisfied for $k = N_s$, since h_{N_s} can only be increased.*

Remark 4.2 *Algorithm 4.1 can be initialized such that Assumption 4.2 is satisfied. Since h_k is an increasing sequence, the condition (4.15) always holds. However, performance of Algorithm 4.1 is sacrificed. Consequently, we do not employ this initialization strategy.*

Theorem 4.1 *Suppose that Assumptions 4.1 and 4.2 are satisfied. Moreover, the maximal velocity, acceleration and jerk are positive, $\mathbf{a}_{max} > \mathbf{0}$, $\mathbf{v}_{max} > \mathbf{0}$, $\mathbf{j}_{max} > \mathbf{0}$, then Algorithm 4.1 always terminates.*

The proof of Theorem 4.1 is given in Appendix A.1.

4.5 Gradient-free Trajectory Optimization

Finally, an optimization algorithm is employed in order to minimize the trajectory total travel time. Since the gradient information $\nabla T(\Theta)$ is not available and can not be analytically computed, derivative-free optimization algorithms are chosen. However, several derivative-free optimization algorithms have been developed and their properties have been well studied [53, 83]. For the sake of simplicity, the resilient back-propagation (PROP) is modified and partially applied for the trajectory optimization. In fact, the PROP step size update rules are deployed, which is well-known for their robust performance and fast convergence [83]. Furthermore, these rules are integrated into the trajectory optimization, which is shown as pseudocodes in Algorithm 4.2 and can be briefly described as follows.

Based on a list of way points $\mathbf{W}$, the initial path parameter Θ_{ini} is obtained by using the heuristics mentioned in Section 4.3. As stated in lines 2-5 of Algorithm 4.2, an initial trajectory is generated by applying Algorithm 4.1. Afterwards, the path parameter Θ is iteratively modified and improved. If a collision between a modified trajectory and the list of obstacles $\mathbf{O}$ is found, the travel time is assigned to infinite. Additionally, a collision checker can be efficiently implemented by utilizing the convex hull property of quintic Bézier splines and applying the separating hyperplane Theorem 3.1. Otherwise, the travel time can be computed for a modified path parameter

Θ_{mod} in analogy to the initial trajectory. On one hand, the new path parameter is assigned and the step size is increased by a factor of 1.2 if the performance is improved. Conversely, the step size is decreased by a factor of 0.5 and the search direction is inverted as shown in lines 17-23 of Algorithm 4.2. Finally, the trajectory optimization is terminated when the planning time is up or the step size is smaller than the minimal threshold $\Delta\Theta_{\mathrm{min}}$.

Algorithm 4.2 Finding the optimal path parameter Θ^*

    Input: The initial path parameter Θ_{ini}, the step size $\Delta\Theta_{\mathrm{ini}}$, the minimal threshold $\Delta\Theta_{\mathrm{min}}$ and a list of obstacles $\mathbf{O}$

    Output: The optimal path parameter Θ^* and the optimal trajectory $\mathbf{q}^*(t)$

    initialization:

1:     $\Theta^* \quad \Theta_{\mathrm{ini}}$;

2:     $\Delta\Theta \quad \Delta\Theta_{\mathrm{ini}}$;

3:     $\mathbf{Q}_{\mathrm{ini}} \quad \mathrm{sampleSmoothPath}(\Theta_{\mathrm{ini}})$;

4:     $\mathbf{h}_{\mathrm{ini}} \quad \mathrm{determineTimeIntervals}(\mathbf{Q}_{\mathrm{ini}})$;

5:     $T^* \quad \sum_{k=1}^{N_{\mathrm{s}}} h_{k,\mathrm{ini}}$;   *// calculate total travel time*

6: **repeat**

7:     **for all** $i = 1 : 1 : 3(N_w - 2)$ **do**

8:        $\Theta_{\mathrm{mod}} \quad \Theta^*$;

9:        $\Theta_{\mathrm{mod},i} \leftarrow \Theta_i^* + \Delta\Theta_i$;

10:      **if** $\mathrm{collision}(\Theta_{\mathrm{mod}}, \mathbf{O})$ **then**

11:         $T_{\mathrm{mod}} \quad \infty$;

12:      **else**

13:         $\mathbf{Q}_{\mathrm{mod}} \quad \mathrm{sampleSmoothPath}(\Theta_{\mathrm{mod}})$;

14:         $\mathbf{h}_{\mathrm{mod}} \quad \mathrm{determineTimeIntervals}(\mathbf{Q}_{\mathrm{mod}})$;

15:         $T_{\mathrm{mod}} \quad \sum_{k=1}^{N_{\mathrm{s}}} h_{k,\mathrm{mod}}$;

16:      **end if**

17:      **if** $(T_{\mathrm{mod}} < T^*)$ **then**

18:         $T^* \quad T_{\mathrm{mod}}$;

19:         $\Theta^* \quad \Theta_{\mathrm{mod}}$;

20:         $\Delta\Theta_i \quad 1.2\Delta\Theta_i$;   *// increase step size*

21:      **else**

22:         $\Delta\Theta_i \quad -0.5\Delta\Theta_i$;   *// invert and decrease step size*

23:      **end if**

24:     **end for**

25: **until** (planning time is up) $\vee$ $(\Delta\Theta \leq \Delta\Theta_{\mathrm{min}})$

4.6   Illustrative Examples

In this section, the efficiency of the proposed approach is demonstrated by two simple scenarios inspired by industrial assembly applications. In these scenarios, the sampling time T_{s} is 5 ms.

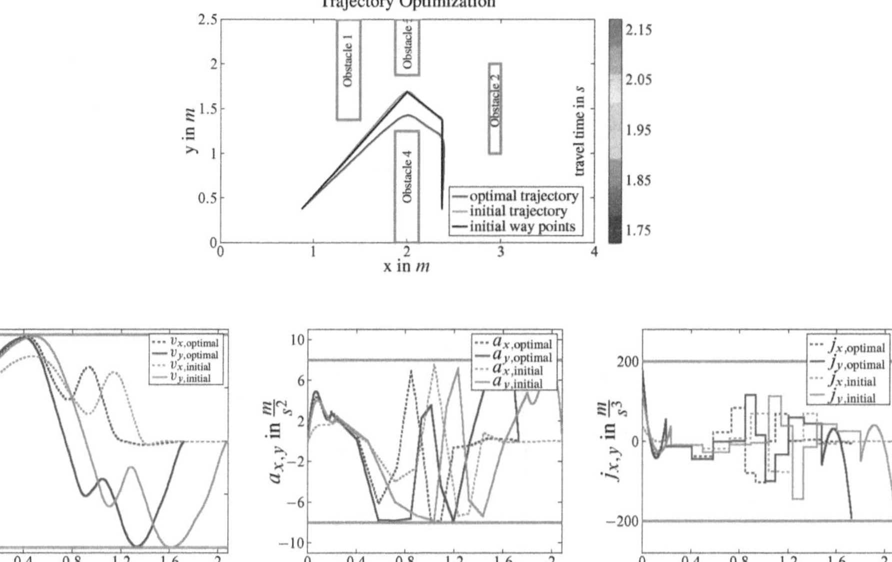

Figure 4.4: The initial velocities and accelerations are respectively $\dot{\mathbf{q}}_{\text{initial}} = \begin{bmatrix} 1 & 1 \end{bmatrix}^T \frac{m}{s}$ and $\ddot{\mathbf{q}}_{\text{initial}} = \begin{bmatrix} 0 & 0 \end{bmatrix}^T \frac{m}{s^2}$. Moreover, all kinematic constraints are satisfied as shown in Figure 4.4, which are identical for both x and y coordinates as $v_{\text{max},\xi} = 2\frac{m}{s}$, $a_{\text{max},\xi} = 8\frac{m}{s^2}$, $j_{\text{max},\xi} = 200\frac{m}{s^3}$, $\xi \in \{x, y\}$.

Scenario 1:

In Scenario 1, the working environment is assumed to be an rectangle, which has a width of 2.5 m and a length of 4 m. Moreover, four obstacles lie in the working environment, whose sizes are correspondingly shown in Figure 4.4. In the experiment, the RRT geometrical planner is used in order to obtain a list of way points, which are respectively $\mathbf{W} := \{\text{start} \equiv \mathbf{w}_1 = [0.875 \quad 0.375]^T, \mathbf{w}_2 = [2.0000 \quad 1.6875]^T, \mathbf{w}_3 = [2.3750 \quad 1.3750]^T, \text{goal} \equiv \mathbf{w}_4 = [2.3750 \quad 0.3750]^T\}$. The travel time of the initial trajectory is $T_{\text{initial}} = 2.085s = 417T_s$, while the travel time of the optimized trajectory is $T_{\text{optimal}} = 1.725s = 345\,T_s$. In other words, performance is improved by almost 20% by employing the trajectory optimization.

Scenario 2:

In Scenario 2, an additional obstacle cost is incorporated into the trajectory optimization algorithm. In order to guarantee safety, the maximal velocity is limited by not only the kinematic constraints but also by distances to the nearest obstacle. Let assume $d_\xi(\mathbf{q})$ is the ξ-distance to the nearest obstacle from the position $\mathbf{q}$, $\xi \in \{x, y\}$. In order to stop the whole system safely, the

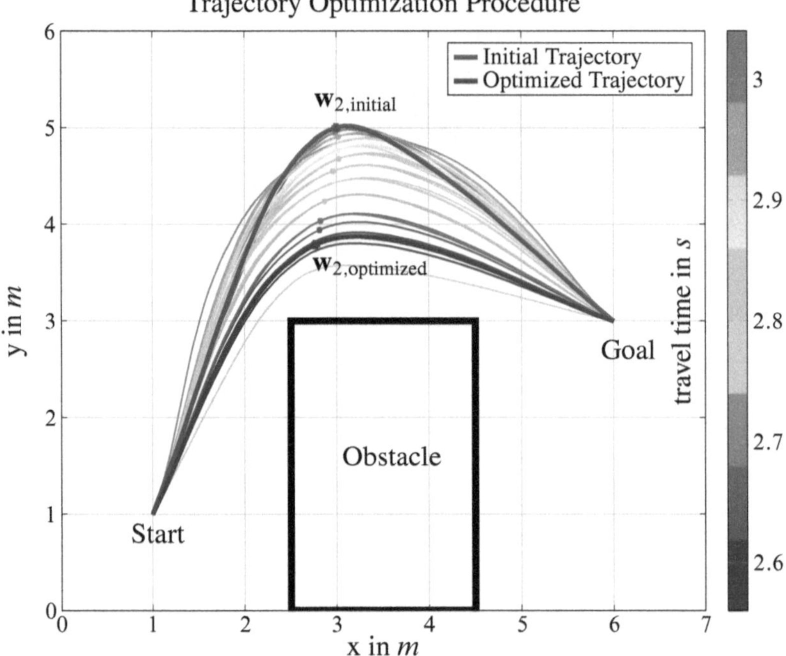

Figure 4.5: A list of collision free way points is given as $\mathbf{W} := \{\text{start} = \mathbf{w}_1 = [1 \quad 1]^T, \mathbf{w}_2 = [3 \quad 5]\}, \mathbf{w}_3 = [6 \quad 3]^T\}$. Moreover, the initial velocities and accelerations are zero as $\dot{\mathbf{q}}_{\text{initial}} = [0 \quad 0]^T \frac{m}{s}$, $\ddot{\mathbf{q}}_{\text{initial}} = [0 \quad 0]^T \frac{m}{s^2}$. All kinematic constraints are identical for both x and y co-ordinates as $v_{\text{max},\xi} = 3\frac{m}{s}$, $a_{\text{max},\xi} = 10\frac{m}{s^2}$ and $j_{\text{max},\xi} = 800\frac{m}{s^3}$. The position of the middle way points during the trajectory optimization is marked as shown in Figure 4.5. Furthermore, the travel time of the initial trajectory is $T_{\text{initial}} = 3.030 = 606T_s$, while the travel time of the optimized trajectory is $T_{\text{optimal}} = 2.560 = 512T_s$. In other words, the overall performance is improved by almost 16%.

maximal allowable velocity $v_{\text{safe,max},\xi}(\mathbf{q})$ can be calculated based on the distance $d_\xi(\mathbf{q})$ as

$$d_\xi(\mathbf{q}) = v_{\text{safe,max},\xi}(\mathbf{q})t_s - \frac{1}{2}a_{\text{max},\xi}t_s^2, \tag{4.16}$$

where the stop time t_s can be defined by

$$t_s = \frac{v_{\text{safe,max},\xi}(\mathbf{q})}{a_{\text{max},\xi}}, \tag{4.17}$$

without considering the jerk constraint. By substituting (4.17) into (4.16), one has

$$d_\xi(\mathbf{q}) = \frac{v_{\text{safe,max},\xi}^2}{2a_{\text{max},\xi}} \Rightarrow v_{\text{safe,max},\xi}(\mathbf{q}) = \sqrt{2d_\xi(\mathbf{q})a_{\text{max},\xi}}.$$

Additionally, the maximal velocity is limited by the kinematic constraint $v_{\max,\xi}$. Therefore, the maximal velocity at the position $\mathbf{q}$ is redefined by

$$
\begin{aligned}
\tilde{v}_{\max,\xi}(\mathbf{q}) &= \min\{v_{\max,\xi}, v_{\text{safe},\max,\xi}(\mathbf{q})\} \\
&= \begin{cases} v_{\text{safe},\max,\xi}(\mathbf{q}), & \text{if } v_{\text{safe},\max,\xi}(\mathbf{q}) = \sqrt{2d_\xi(\mathbf{q})a_{\max,\xi}} \leq v_{\max,\xi}, \\ v_{\max,\xi}, & \text{if } v_{\text{safe},\max,\xi}(\mathbf{q}) = \sqrt{2d_\xi(\mathbf{q})a_{\max,\xi}} > v_{\max,\xi}. \end{cases}
\end{aligned}
$$

As as a consequence, the maximal velocity on a segment S_k between two interpolating points $\mathbf{q}_k$ and $\mathbf{q}_{k+1}$ is set as

$$
\tilde{v}_{\max,\xi,S_k} = \min\{\tilde{v}_{\max,\xi}(\mathbf{q}_k), \tilde{v}_{\max,\xi}(\mathbf{q}_{k+1})\}.
$$

By doing so, the additional obstacle cost can be indirectly integrated into the trajectory optimization. The optimization procedure of the middle way point is depicted in Figure 4.5.

4.7 Conclusion

In this chapter, we presented a novel decompositional approach to generate high-order near time-optimal trajectories. Moreover, the ingredient algorithms are described in detail. Furthermore, the termination analysis and the proof of the decompositional approach are provided. Additionally, the effectiveness and successful application for two-dimensional robots are demonstrated by illustrative examples.

In order to classify our results, a short comparison to cutting edge trajectory optimization approaches is given in the following. Particularly, the time-optimal trajectory generation approach is presented for omnidirectional robots [91, 93]. However, the jerk constraint as well as the time discretization effect are not considered in contrast to our approach. Another alternative leading tool is Covariant Hamiltonian Optimization for Motion Planning (CHOMP), which relies on an effective functional gradient method [113]. Nevertheless, a time-gridding way-point representation is needed for numerical optimization by CHOMP, which leads to a trajectory optimization problem with more numerous variables than our proposed approach. Moreover, time-optimal trajectories can be generated by adopting B-spline representation [67, 111]. Based on cubic B-splines, nonlinear optimization is employed to find optimal time sequences [111]. However, the main disadvantage of this approach is high computational burden. Furthermore, control points are not optimized, which can lead to a poor performance solution. Contrarily, the control points are optimized in the work [67]. Nevertheless, the jerk constraint as well as the time discretization effect are not taken into account by this work. Therefore, we expect that our approach will outperform all aforementioned proposed solutions.

Future work will be devoted to real applications of the proposed approach to two-dimensional robots in order to make an industrial impact. In order to skip local minima and quickly converge to a global strong solution, another improvement perspective is employing practical effective variants on the robust optimization algorithm in lieu of the simple RPROP update rules. Finally, our approach can be simply extended and applied to other robot systems.

5 A Centralized Coordination Solution for Multirobot Systems

In this chapter, a centralized coordination solution for the multirobot systems is introduced, which is composed of a collision avoidance algorithm and a conflict resolution algorithm. Particularly, the collision avoidance algorithm is able to guarantee safety invariantly, while the conflict resolution algorithm is applied to prevent deadlocks such that each robot can reach its goal. Moreover, effectiveness, low computational complexity as well as performance of the centralized coordination solution are demonstrated in simulations. Finally, the chapter is concluded with a discussion about potential applications.

5.1 Introduction

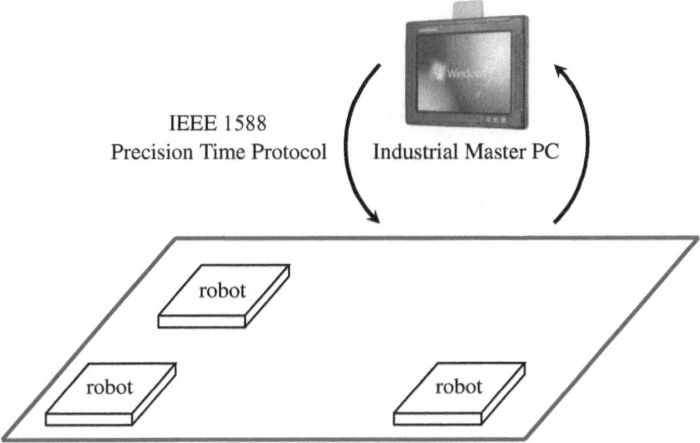

Figure 5.1: The centralized system architecture

Firstly, the centralized system architecture is recapped. As shown in Figure 5.1 and mentioned in Chapter 1, the whole system consists of a master PC and multiple robots. This centralized system architecture is prevalent in several commercial products [8, 11]. In this centralized system architecture, all decisions are made by the master PC. In other words, the master PC is responsible for generating all robot trajectories as well as executing the coordination solution. Although, all robot states and other necessary information are accessible by the master PC, it is still very challenging to develop a centralized coordination solution.

In order to be highly modular, the centralized coordination solution should operate as a post process to the trajectory optimization presented in Chapter 4. Consequently, the kinematic constraints such as maximal velocity, acceleration and jerk as well as the time discretization consideration can be preserved for the purpose of obtaining high performance motion control systems [54]. As aforementioned, the centralized coordination solution is composed of a avoidance algorithm and a conflict resolution algorithm. In the following, the working principles of these algorithms as well as their fundamental ingredients are explained in detail. Finally, this chapter is structured by the following sections:

- **A Centralized Collision Avoidance Algorithm:** The centralized collision avoidance algorithm is described, which consists of a high-order braking trajectory generation algorithm and an efficient continuous collision detection algorithm.

- **A High-order Braking Trajectory Generation Algorithm:** The high-order braking trajectory generation algorithm is presented.

- **An Efficient Continuous Collision Detection Algorithm:** In order to detect mutual collisions among robots, a tunneling effect must be taken into account. In this section, this tunneling effect is elucidated. Subsequently, the continuous collision detection algorithm is introduced for the purpose of preventing the tunneling effect.

- **An Invariant Safety Guarantee Proof:** A formal safety proof of the centralized collision avoidance algorithm is given.

- **A Conflict Resolution Algorithm:** The conflict resolution algorithm as well as its components are explained.

- **Algorithms Simulation:** Simulations are used to validate efficiency of the centralized coordination solution.

- **Conclusion:** Lastly, some important remarks are drawn to conclude the chapter.

5.2   A Centralized Collision Avoidance Algorithm

In this section, a centralized collision avoidance algorithm is proposed. Moreover, the algorithm can be scaled with a moderate number of robots. Particularly, the algorithm is similar to the dynamical safety search (DSS), which is a well-known centralized multirobot motion planning approach [11, 12]. According to our knowledge, the DSS is the only centralized approach, which is successfully employed in practical applications [11]. Firstly, the working principle and basic properties of the DSS are introduced.

The DSS was developed by James Bruce during world robot soccer competitions [11], which is based on the Fox's single agent dynamical window approach (DWA) [30]. The DSS extends the DWA to solve the multirobot motion planning problem. In fact, a grid-based search approach on a velocity space in the DWA is replaced by a randomized algorithm on a two dimensional acceleration space in the DSS. Moreover, the DSS can be deployed as an any-time algorithm, which is a paramount property for real time motion planning. Basically, the invariant safety can be guaranteed by the DSS as well as the DWA. Furthermore, the DSS has a polynomial computational complexity. In fact, the computational burden of the DSS increases quadratically with the number of robots. Nevertheless, the DSS lacks the completeness in contrast to the multiple joint planning algorithm, which has high computational burden as mentioned in Chapter 2. As a consequence, the multiple joint planning algorithm is impractical for real applications. In the DSS, a random search procedure is performed on the acceleration space. Consequently, high-order trajectories can not be generated by the DSS, since the acceleration is discontinuous. Additionally, the DSS approach is not able to deliver a deterministic consistent performance, which is desired for some industrial applications [8]. By slightly modifying the DSS, all aforementioned shortcomings can be mitigated by the proposed centralized collision avoidance algorithm.

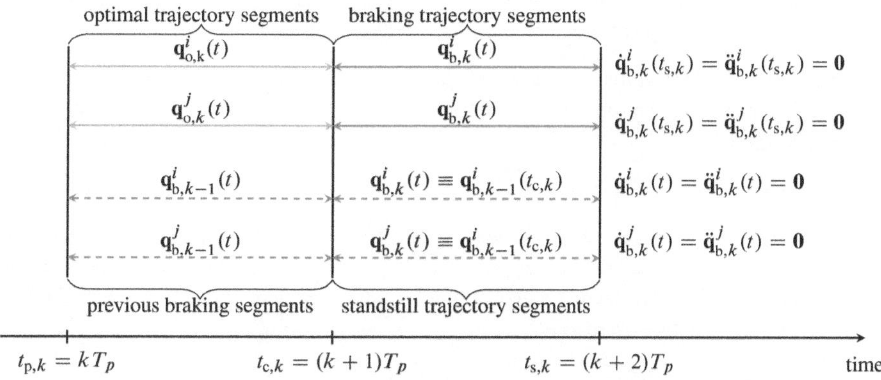

Figure 5.2: The working principle of the centralized collision avoidance algorithm

For the sake of simplifying explanation, any two robots i and j are considered. Let T_p define the planning cycle of the centralized collision avoidance algorithm. Particularly, the cycle T_p is an integer number of the sampling time T_s, i.e. $T_\mathrm{p} = n_\mathrm{p} T_\mathrm{s}$, $n_\mathrm{p} \in \mathbb{N}$. In each planning cycle $t_{\mathrm{p},k} = k T_\mathrm{p}$, $k \in \mathbb{N}$, the collision avoidance algorithm is divided into two steps. In the first step, an optimal trajectory segment $\mathbf{q}_{\mathrm{o},k}^i(t)$ over the time interval $[t_{\mathrm{p},k} = k T_\mathrm{p} \quad t_{\mathrm{c},k} = (k+1) T_\mathrm{p}]$ for each robot i is generated by the trajectory optimization algorithm presented in Chapter 4. As shown in Figure 5.2, the optimal trajectory segment is followed by a braking trajectory segment $\mathbf{q}_{\mathrm{b},k}^i(t)$ in the time interval $[t_{c,k} = (k+1)T_p \quad t_{s,k} = (k+2)T_p]$, which brings the robot i to a standstill state, where $\dot{\mathbf{q}}^i_{b,k}(t_{s,k}) = \ddot{\mathbf{q}}^i_{b,k}(t_{s,k}) = \mathbf{0}$. In the second step, the executed trajectory segment is chosen for any robot i based on the following three cases:

Case 1:

The concatenation of the optimal and braking trajectory segments $\mathbf{q}^i_{p,k} = [\mathbf{q}^i_{o,k}(t) \quad \mathbf{q}^i_{b,k}(t)]$ of the robot i does not collide with that $\mathbf{q}^j_{p,k} = [\mathbf{q}^j_{o,k}(t) \quad \mathbf{q}^j_{b,k}(t)]$ of the robot j. Consequently, the optimal trajectory segment $\mathbf{q}^i_{o,k}(t)$ for the robot i is selected in the next planning period. **Case 1** is ideal, since the optimal trajectory segments $\mathbf{q}^i_{o,k}(t)$ and $\mathbf{q}^j_{o,k}(t)$ can be followed by both robots.

Case 2:

In contrast to **Case 1**, a collision exists between the concatenations $\mathbf{q}^i_{p,k} = [\mathbf{q}^i_{o,k}(t) \quad \mathbf{q}^i_{b,k}(t)]$ and $\mathbf{q}^j_{p,k} = [\mathbf{q}^j_{o,k}(t) \quad \mathbf{q}^j_{b,k}(t)]$. Moreover, let's suppose that the robot i has a higher priority than the robot j. In this case, the continuous collision detection algorithm is applied for the concatenation $[\mathbf{q}^i_{o,k}(t) \quad \mathbf{q}^i_{b,k}(t)]$ of the robot i and the previous braking trajectory segment $\mathbf{q}^j_{b,k-1}(t)$ of the robot j. Based on the result of this continuous collision detection, **Case 2** is split into the following subcases:

- **Subcase 2.1:** If no collision is found, the optimal trajectory segment $\mathbf{q}^i_{o,k}(t)$ for the robot i can be executed in the next planning cycle, i.e. $\mathbf{q}^i_{p,k} = [\mathbf{q}^i_{o,k}(t) \quad \mathbf{q}^i_{b,k}(t)]$.

- **Subcase 2.2:** If any collision is detected, the previous braking trajectory segment $\mathbf{q}^i_{b,k-1}(t))$ will be chosen for the robot i in the next planning cycle as $\mathbf{q}^i_{p,k} = [\mathbf{q}^i_{b,k-1}(t) \quad \mathbf{q}^i_{b,k}(t) = \mathbf{q}^i_{b,k-1}(t_{c,k})]$.

Case 3:

In contrast to **Case 1** and **Case 2**, collisions are detected between the trajectory concatenations $\mathbf{q}^i_{p,k} = [\mathbf{q}^i_{o,k}(t) \quad \mathbf{q}^i_{b,k}(t))]$ and $\mathbf{q}^j_{p,k} = [\mathbf{q}^j_{o,k}(t) \quad \mathbf{q}^j_{b,k}(t)]$ as well as the robot i has a lower priority than the robot j. In this case, the previous braking trajectory segment $\mathbf{q}^i_{b,k-1}(t)$ is selected for the robot i in the next planning cycle as $\mathbf{q}^i_{p,k} = [\mathbf{q}^i_{b,k-1}(t) \quad \mathbf{q}^i_{b,k}(t) = \mathbf{q}^i_{b,k-1}(t_{c,k})]$.

In general, the centralized collision avoidance algorithm can be adopted for a set of N_r robots. In fact, the guidelines are applied for the robot i with respect to any other robot j. Moreover, if the previous braking trajectory segment $\mathbf{q}^i_{b,k-1}(t)$ for the robot i is selected, than no continuation is necessary. Consequently, the collision avoidance algorithm can be terminated for the robot i. Furthermore, it is assumed that any robot can be braked to a standstill state within one planning cycle T_p as shown in Figure 5.2 in order to simplify the explanation. Nevertheless, the collision avoidance algorithm can be extended to the case, when the braking time $t_{b,k} = t_{s,k} - t_{c,k}$ is larger than one planning cycle T_p. Obviously, the braking trajectory generation algorithm and the continuous collision detection algorithm are the most important ingredients, which will be described in the two following sections.

5.3 A High-order Braking Trajectory Generation Algorithm

In this section, a high-order braking trajectory generation algorithm is introduced. Basically, the high-order braking trajectories should satisfy boundary conditions, which are the initial and end states. The initial state expresses respectively the position $\mathbf{q}(t_c)$, the velocity $\dot{\mathbf{q}}(t_c)$, and the acceleration $\ddot{\mathbf{q}}(t_c)$ at the time stamp t_c. The end state defines the position $\mathbf{q}(t_s)$, the end velocity $\dot{\mathbf{q}}(t_s)$ and the acceleration $\ddot{\mathbf{q}}(t_s)$, when the robots are stopped at the time stamp t_s. Obviously, the end position $\mathbf{q}(t_s)$ can be flexible, while the end velocity and acceleration are zero. Ultimately, the following boundary conditions

$$\mathbf{q}_b(t_c) = \mathbf{q}(t_c), \quad \dot{\mathbf{q}}_b(t_c) = \dot{\mathbf{q}}(t_c), \quad \ddot{\mathbf{q}}_b(t_c) = \ddot{\mathbf{q}}(t_c), \quad \dot{\mathbf{q}}_b(t_s) = \mathbf{0} \quad \text{and} \quad \ddot{\mathbf{q}}_b(t_s) = \mathbf{0}. \tag{5.1}$$

must be fulfilled by a braking trajectory segment $\mathbf{q}_b(t)$. Additionally, the following kinematic constraints

$$|\dot{\mathbf{q}}_b(t)| \leq \mathbf{v}_{\max}, \quad |\ddot{\mathbf{q}}_b(t)| \leq \mathbf{a}_{\max} \quad \text{and} \quad |\dddot{\mathbf{q}}_b(t)| \leq \mathbf{j}_{\max}. \tag{5.2}$$

have to be satisfied. Furthermore, the aforementioned discretization effect should be considered in the braking trajectory segment. In other words, the jerk $\dddot{\mathbf{q}}_b(t)$ can only be discontinuous at a time stamp $t_d = n_d T_s, n_d \in \mathbb{N}$, which is an integer of the sampling time T_s. Moreover, the braking trajectory segment $q_{b,\xi}(t)$ in each dimension $\xi \in \{x, y\}$ can be independently generated due to the robot holonomic constraints. Since the five boundary conditions (5.1) and the maximal jerk constraint must be taken into account, a natural idea is applying a fourth degree polynomial to generate the braking trajectory segment as follows:

$$q_{b,\xi}(t) = b_4 t^4 + b_3 t^3 + b_2 t^2 + b_1 t + b_0, \quad t \in [t_c \quad t_s = t_c + T_b],$$

where T_b is the duration. From (5.1), the coefficients $b_i, i \in \{0, 1, 2, 3, 4\}$, can be computed if the duration T_b is given. If the kinematic constraints (5.2) are not satisfied, the duration will be increased by T_s. Since the braking trajectory segment is a fourth degree polynomial, the maximal velocity, acceleration and jerk can be computed analytically. However, a main drawback of this approach is that the duration T_b is not minimal. Furthermore, a feasible solution can occasionally not be found. Consequently, an alternative approach is proposed, which is able to overcome these drawbacks. In this approach, the braking trajectory segment $q_{b,\xi}(t)$ is split into either three or four cubic polynomials and can be described by the following cases:

Case 1: $\dot{q}_{b,\xi}(t_c)\ddot{q}_{b,\xi}(t_c) \leq 0$
The braking trajectory segment is divided into three piecewise cubic polynomials and normalized by the time variable τ as follows:

$$\begin{aligned}
q_{1,b,\xi}(\tau) &= b_{1,3}\tau^3 + b_{1,2}\tau^2 + b_{1,1}\tau + b_{1,0}, & \tau &= t - t_c, & \tau &\in [0 \quad T_1], \\
q_{2,b,\xi}(\tau) &= b_{2,3}\tau^3 + b_{2,2}\tau^2 + b_{2,1}\tau + b_{2,0}, & \tau &= t - t_c - T_1, & \tau &\in [0 \quad T_2], \\
q_{3,b,\xi}(\tau) &= b_{3,3}\tau^3 + b_{3,2}\tau^2 + b_{3,1}\tau + b_{3,0}, & \tau &= t - t_c - T_1 - T_2, & \tau &\in [0 \quad T_3],
\end{aligned} \tag{5.3}$$

where $b_{i,n}$ defines the coefficients of the ith cubic polynomial and T_1, T_2, T_3 are respectively the unknown time intervals. Moreover, the second polynomial is assumed to have a constant acceleration, which leads to $b_{2,3} = 0$. From (5.1), the position, velocity and acceleration at the time stamp

t_c is predefined, one consequently has

$$b_{1,0} = q_\xi(t_c), \quad b_{1,1} = \dot{q}_\xi(t_c) \quad \text{and} \quad b_{1,2} = \frac{\ddot{q}_\xi(t_c)}{2}.$$

The remainder unknown variables are defined by

$$\mathbf{b} = \begin{bmatrix} b_{1,3} & b_{2,0} & b_{2,1} & b_{2,2} & b_{3,0} & b_{3,1} & b_{3,2} & b_{3,3} \end{bmatrix}^T.$$

Furthermore, the continuous position, velocity and acceleration at the time stamps $t_c + T_1$ and $t_c + T_1 + T_2$ lead to

$$\begin{aligned}
T_1^3 b_{1,3} - b_{2,0} &= -b_{1,0} - b_{1,1}T_1 - b_{1,2}T_1^2, \\
3T_1^2 b_{1,3} - b_{2,1} &= -b_{1,1} - 2b_{1,2}T_1, \\
6T_1 b_{1,3} - 2b_{2,2} &= -2b_{1,2},
\end{aligned} \tag{5.4}$$

and

$$\begin{aligned}
b_{2,0} + T_2 b_{2,1} + T_2^2 b_{2,2} - b_{3,0} &= 0, \\
b_{2,1} + 2T_2 b_{2,2} - b_{3,1} &= 0, \\
2b_{2,2} - 2b_{3,2} &= 0.
\end{aligned} \tag{5.5}$$

Moreover, the boundary conditions $\dot{q}_{b,\xi}(t_s) = 0$ and $\ddot{q}_{b,\xi}(t_s) = 0$ with $t_s = t_c + T_1 + T_2 + T_3$ is fulfilled by

$$\begin{aligned}
b_{3,1} + 2T_3 b_{3,2} + 3T_3^2 b_{3,3} &= 0, \\
2b_{3,2} + 6T_3 b_{3,3} &= 0.
\end{aligned} \tag{5.6}$$

By summarizing (5.4), (5.5) and (5.6), one obtains that

$$\underbrace{\begin{bmatrix}
T_1^3 & -1 & 0 & 0 & 0 & 0 & 0 & 0 \\
3T_1^2 & 0 & -1 & 0 & 0 & 0 & 0 & 0 \\
6T_1 & 0 & -2 & 0 & 0 & 0 & 0 & 0 \\
0 & 1 & T_2 & T_2^2 & -1 & 0 & 0 & 0 \\
0 & 0 & 1 & 2T_2 & 0 & -1 & 0 & 0 \\
0 & 0 & 0 & 1 & 0 & 0 & -1 & 0 \\
0 & 0 & 0 & 0 & 0 & 1 & 2T_3 & 3T_3^2 \\
0 & 0 & 0 & 0 & 0 & 0 & 2 & 6T_3
\end{bmatrix}}_{A_{b_\xi}}
\underbrace{\begin{bmatrix}
b_{1,3} \\ b_{2,0} \\ b_{2,1} \\ b_{2,2} \\ b_{3,0} \\ b_{3,1} \\ b_{3,2} \\ b_{3,3}
\end{bmatrix}}_{\mathbf{b}_\xi}
=
\underbrace{\begin{bmatrix}
-q_\xi(t_c) - \dot{q}_\xi(t_c)T_1 - \frac{1}{2}\ddot{q}_\xi(t_c)T_1^2 \\
-\dot{q}_\xi(t_c) - \ddot{q}_\xi(t_c)T_1 \\
-\ddot{q}_\xi(t_c) \\
0 \\
0 \\
0 \\
0 \\
0
\end{bmatrix}}_{\mathbf{v}_\xi}$$

$$\Rightarrow \mathbf{b}_\xi = A_{b_\xi}^{-1}\mathbf{v}_\xi, \tag{5.7}$$

which implies that the spline coefficients can be obtained if the time intervals T_1, T_2 and T_3 are given. By imitating the idea of the near time-optimal trajectory fitting presented in Chapter 4, the following braking trajectory generation algorithm is implemented by using pseudocode shown in Algorithm 5.1. First, the time intervals are initialized as

$$T_1 = \lceil \frac{a_{\max} - |\ddot{q}_\xi(t_c)|}{j_{\max}T_s} \rceil T_s, \quad T_2 = \lceil \frac{|\dot{q}_\xi(t_c)|}{a_{\max}T_s} \rceil T_s, \quad T_3 = \lceil \frac{|a_{\max}|}{j_{\max}T_s} \rceil T_s, \tag{5.8}$$

where $\lceil \epsilon \rceil$ denotes the least integer greater than ϵ. Afterwards, the time interval T_i, $i \in \{1, 2, 3\}$, will be increased by T_s, if any kinematic constraint is not fulfilled in the polynomial. Since the braking trajectory polynomials are cubic, the kinematic constraints can be fast and analytically verified by using closed form expressions.

Algorithm 5.1 Computing the braking trajectory $q_{b,\xi}(t)$

    Input: The initial state $q_\xi(t_c)$, $\dot{q}_\xi(t_c)$ and $\ddot{q}_\xi(t_c)$
    Output: The braking trajectory $q_{b,\xi}(t)$
1: **Initialization:**
2:     $T_1 = \lceil \frac{a_{\max} - |\ddot{q}_\xi(t_c)|}{j_{\max} T_s} \rceil T_s; \quad T_2 = \lceil \frac{|\dot{q}_\xi(t_c)|}{a_{\max} T_s} \rceil T_s; \quad T_3 = \lceil \frac{|a_{\max}|}{j_{\max} T_s} \rceil T_s;$
3:     $\mathbf{b}_\xi = \mathbf{A}_{b_\xi}^{-1} \mathbf{v}_\xi$
4: **repeat**
5:     **for all** $k = 1 : 1 : 3$ **do**
6:         **if** $\left((\max |\dot{q}_{k,b,\xi}(t)| > v_{\max}) \vee (\max |\ddot{q}_{k,b,\xi}(t)| > a_{\max}) \vee (\max |\dddot{q}_{k,b,\xi}(t)| > j_{\max}) \right)$ **then**
7:             $T_k \quad T_k + T_s;$
8:         **end if**
9:     **end for**
10:     $\mathbf{b}_\xi = \mathbf{A}_{b_\xi}^{-1} \mathbf{v}_\xi$
11: **until** $(\max |\dot{q}_{b,\xi}(t)| \le v_{\max}) \wedge (\max |\ddot{q}_{b,\xi}(t)| \le a_{\max}) \wedge (\max |\dddot{q}_{b,\xi}(t)| \le j_{\max})$

Case 2: $\dot{q}_{b,\xi}(t_c) \ddot{q}_{b,\xi}(t_c) > 0$

The initial velocity and acceleration has same signs. An additional cubic polynomial is incorporated into the braking trajectory segment, which reverses the acceleration sign in the shortest time. This polynomial can be defined as

$$q_{b,\xi}(\tau) = q_{b,\xi}(t_c) + \dot{q}_{b,\xi}(t_c)\tau + \frac{1}{2}\ddot{q}_{b,\xi}(t_c)\tau^2 + \frac{1}{6}j_{b,a,\xi}\tau^3, \quad \tau = t - t_c, \quad \tau \in [0 \quad T_a],$$

where

$$T_a = \lceil \frac{|\ddot{q}_{b,\xi}(t_c)|}{j_{\max} T_s} \rceil T_s \quad \text{and} \quad j_{b,a,\xi} = -\frac{\ddot{q}_{b,\xi}(t_c)}{T_a}. \tag{5.9}$$

At the time stamp $t_c + T_a$, the acceleration $\ddot{q}_{b,\xi}(t_c + T_a)$ is zero. Consequently, Algorithm 5.1 can be applied to compute the remainder of the braking trajectory segment $q_{b,\xi}(t)$.

5.4 An Efficient Continuous Collision Detection Algorithm

Let assume that the robot 1 is hold while the robot 2 translates during the time interval kT_s and $(k+1)T_s$ as shown in Figure 5.3, where T_s is the sampling time and $k \in \mathbb{N}$ is an integer. Obviously, no intersection is detected between the two robots at the discrete time instants kT_s and $(k+1)T_s$. Nevertheless, a collision can occur between the two time instants. This phenomenon is called the tunneling effect [64]. In order to prevent this tunneling effect, a continuous collision detection algorithm is employed among robots. Generally, continuous collision detection algorithms

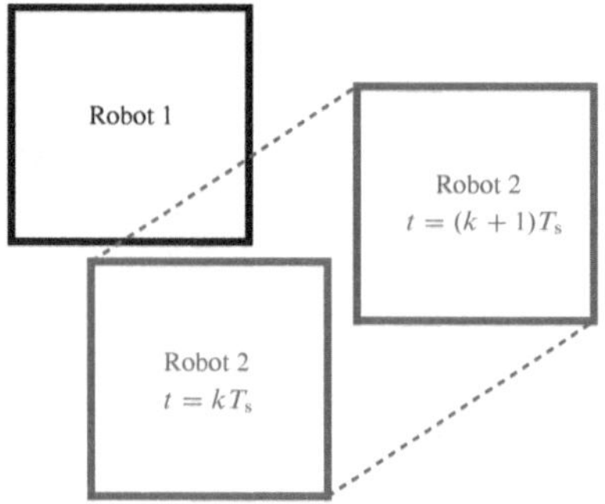

Figure 5.3: The tunneling effect

require more computational burden than discrete ones [71, 79, 90, 97]. However, a fast continuous collision detection algorithm can be developed by utilizing robot trajectory and geometrical properties. In the following, the working principle of the continuous collision detection algorithm is explained.

Particularly, the robot geometric shapes are squares. Therefore, a robot i collides with a robot j, if and only if there exists at least a time interval $\mathbb{T}_{\text{col}}$, in which x- and y- distances between the two robots are simultaneously smaller than the robot width L. In other words, one has

$$|l_x| = |q_x^i(t) - q_x^j(t)| < L \quad \wedge \quad |l_y| = |q_y^i(t) - q_y^j(t)| < L, \quad \forall t \in \mathbb{T}_{\text{col}}, \tag{5.10}$$

where $q_\xi^{i,j}(t)$, $\xi \in \{x, y\}$, are respectively the x- and y- coordinates of the robots i and j at the time instant t.

Based on this idea, the continuous collision detection algorithm is developed. In fact, it verifies if a time interval $\mathbb{T}_{\text{col}}$ exists. To be more precisely, the continuous collision detection algorithm is constituted by the two following steps:

- **Step 1:** Compute the time intervals

$$\begin{aligned}
\mathbb{I}_x &= \{[t_{1,\text{enter},x} \quad t_{1,\text{leave},x}], \quad [t_{2,\text{enter},x} \quad t_{2,\text{leave},x}], \quad \ldots, \quad [t_{N_x,\text{enter},x} \quad t_{N_x,\text{leave},x}]\}, \\
\mathbb{I}_y &= \{[t_{1,\text{enter},y} \quad t_{1,\text{leave},y}], \quad [t_{2,\text{enter},y} \quad t_{2,\text{leave},y}], \quad \ldots, \quad [t_{N_y,\text{enter},y} \quad t_{N_y,\text{leave},y}]\},
\end{aligned} \tag{5.11}$$

where $t_{i,\text{enter},\xi}$ and $t_{i,\text{leave},\xi}$ denotes the time stamps, when the distance l_ξ enters or leaves the interval $[-L \quad L]$ as illustrated in Figure 5.4.

- **Step 2:** Finally, a test is carried out whether there is any overlap between the two time intervals $\mathbb{I}_x$ and $\mathbb{I}_y$, i.e. if $\mathbb{T}_{\text{col}} = \mathbb{I}_x \cap \mathbb{I}_y = \varnothing$. If such an interval exists, collisions between the two robots are detected.

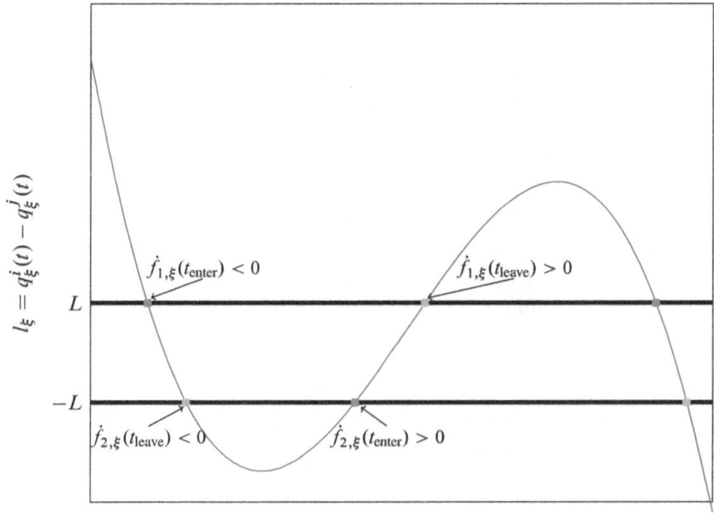

Figure 5.4: Computing the time interval $\mathbb{I}_\xi$, $\xi \in \{x, y\}$. The red dots denote the time stamps when the distance $l_\xi(t)$ enters the interval $[-L \quad L]$. On the other hand, the green dots denotes the time stamps when the distance $l_\xi(t)$ leaves the interval $[-L \quad L]$.

Apparently, **Step 2** can be easily implemented. Moreover, **Step 1** can also be efficiently computed by applying the following trick. Since the robot trajectories in each dimension ξ are piecewise either cubic or quintic polynomials from (4.8), (4.9) and (5.3), the time intervals can be easily computed. Let determine the time stamps, when the distance $l_\xi(t)$ leaves or enters the distance interval from the above margin. By solving the equation

$$f_{1,\xi}(t) = l_\xi(t) - L = q^i_\xi(t) - q^j_\xi(t) - L = c_{n,\xi}t^n + c_{n-1,\xi}t^{n-1} + \cdots + c_{1,\xi}t + c_{0,\xi} = 0, \quad (5.12)$$

where $q^i_\xi(t)$ is defined by (4.8), (4.9) or (5.3) and $n \in \{3, 5\}$, a list of the time stamps $\mathbb{T}_{1,\xi}$ is obtained. Assume τ_1 is an element of the list $\mathbb{T}_{1,\xi}$ and $\dot{f}_{1,\xi}(t) = \frac{df_{1,\xi}}{dt}(t)$ is the first time derivative. Obviously, if $\dot{f}_{1,\xi}(\tau_1) < 0$, then τ_1 is the time stamp, when the distance $l_\xi(t)$ enters the interval $[-L \quad L]$ from the above margin. Conversely, if $\dot{f}_{1,\xi}(\tau_1) > 0$, then τ_1 is the time stamp, when the distance $l_\xi(t)$ leaves the interval $[-L \quad L]$ as shown in Figure 5.4.

Analogously, the distance $l_\xi(t)$ leaves and enters the interval $[-L \quad L]$ from the below margin at a list of time stamps $\mathbb{T}_2$. These time stamps are obtained by solving the equation

$$f_{2,\xi}(t) = l_\xi(t) + L = q_\xi^i(t) - q_\xi^j(t) + L = c_{n,\xi}t^n + c_{n-1,\xi}t^{n-1} + \cdots + c_{1,\xi}t + c_{0,\xi} = 0. \quad (5.13)$$

In contrast to the previous case, if $\tau_2 \in \mathbb{T}_2$ and $\dot{f}_{2,\xi}(\tau_2) > 0$, then τ_2 is the time-stamp, when the distance $l_\xi(t)$ enters the interval $[-L \quad L]$ and vice versus. By sorting the entering and leaving time stamps $\mathbb{T}_{1,\xi}$ and $\mathbb{T}_{2,\xi}$, one has the time interval $\mathbb{I}_\xi$ in **Step 1**.

In most cases, the polynomials $f_{1,\xi}(t)$ and $f_{2,\xi}(t)$ are cubic. Consequently, the equations (5.12) and (5.13) can be solved by trigonometric closed form expressions provided in Appendix A.3. Obviously, only real solutions of these equations are considered. As described in Section 5.2, the continuous collision avoidance algorithm examines only the time interval $\mathbb{T}_{\text{in}} = [t_{\text{p},k} \quad t_{\text{s},k}]$. Therefore, an additional condition is checked in order to find out whether an intersection between the $\mathbb{T}_{\text{col}}$ and $\mathbb{T}_{\text{in}}$ exists. If the intersection is not empty, mutual collisions are detected.

As mentioned in Chapter 3, the broad phase checking can be employed to increase the efficiency of the collision detection algorithm. Roughly speaking, if distances between two robots are too large or robots are placed on distinct regions, mutual collisions can be discarded. Nevertheless, the broad phase checking is not integrated into the continuous collision detection algorithm for the sake of facilitating the implementation.

5.5 An Invariant Safety Guarantee Proof

In order to show that the centralized collision avoidance algorithm is safe, a formal formulation is required. In fact, the collision-free condition can by represented by

$$S(t) = O(\mathbf{q}_{\text{e}}^i(t)) \cap O(\mathbf{q}_{\text{e}}^j(t)) = \varnothing, \quad \forall i, j \in \{1, ..., N_{\text{r}}\}, \quad i \neq j, \quad \forall t, \quad (5.14)$$

where

- $\mathbf{q}_{\text{e}}^i(t)$ and $\mathbf{q}_{\text{e}}^j(t)$ are respectively the executed trajectories of the robots i and j,

- $O(\mathbf{q}_{\text{e}}^i(t))$ and $O(\mathbf{q}_{\text{e}}^j(t))$ are the occupied areas of the robots i and j at the time instant t,

- N_{r} is the number of robots.

In other words, the safety can be invariantly guaranteed if the collision-free condition $S(t) = \varnothing$ holds for all time.

Let's assume that the centralized collision avoidance algorithm is started at the planning cycle $k = 0$ with the following initial condition

$$C^{i,j}(k = 0) = O(\mathbf{q}^i_{\mathrm{p},k}(t)) \cap O(\mathbf{q}^j_{\mathrm{p},k}(t)) = \varnothing,$$
$$\forall\ i,\ j \in \{1, ..., N_{\mathrm{r}}\}, \quad i \neq j, \tag{5.15}$$
$$\forall\ t_{\mathrm{p},k} \leq t \leq t_{\mathrm{s},k},$$

where

$$\mathbf{q}^{\{i,j\}}_{\mathrm{p},k}(t) = [\ \underbrace{\mathbf{q}^{i,j}_{\mathrm{e},k}(t)}_{t_{\mathrm{p},k} \leq t \leq t_{\mathrm{c},k}} \quad \underbrace{\mathbf{q}^{i,j}_{\mathrm{b},k}(t)}_{t_{\mathrm{c},k} \leq t \leq t_{\mathrm{s},k}}\ ]$$

is the concatenation of the two trajectory segments. As shown in Figure 5.2, $\mathbf{q}^{i,j}_{\mathrm{e},k}(t)$ is the executed trajectory segment by the robot i or j, while $\mathbf{q}^{i,j}_{\mathrm{b},k}(t)$ is the subsequent braking trajectory segment. Particularly, $\mathbf{q}^{i,j}_{\mathrm{e},k}(t)$ can be either the optimal trajectory segment $\mathbf{q}^{i,j}_{\mathrm{o},k}(t)$ from the planning cycle k or the braking trajectory segment $\mathbf{q}^{i,j}_{\mathrm{b},k-1}(t)$ from the previous planning cycle $k - 1$. Eventually, one is able to show that if the initial condition (5.15) is satisfied, then the collision avoidance algorithm is invariantly safe by proving the following Theorem.

Theorem 5.1 *For N_r robots, if the condition (5.15) is satisfied, then the centralized collision avoidance algorithm presented in Section 5.2 is safe, i.e. the collision free condition (5.14) holds.*

Proof Firstly, we prove that the trajectories $\mathbf{q}^{i,j}_{\mathrm{p}}(t)$ can always be safely generated by using the collision avoidance algorithm. In other words, the condition

$$C^{i,j}(k) = \quad O(\mathbf{q}^i_{\mathrm{p},k}(t)) \cap O(\mathbf{q}^j_{\mathrm{p},k}(t)) = \varnothing,$$
$$\forall\ i,\ j \in \{1, ..., N_{\mathrm{r}}\}, \quad i \neq j, \tag{5.16}$$
$$\forall\ t_{\mathrm{p},k} \leq t \leq t_{\mathrm{s},k},$$

holds for every planning cycle $k \geq 0$, which can be demonstrated by applying the induction principle. Let assume that $C^{i,j}(k) = \varnothing$, one has to show that $C^{i,j}(k + 1) = \varnothing$. As described in Section 5.2, one of the following cases will be executed in the planning step $k + 1$:

Case 1 or Subcase 2.1:
The optimal trajectory for the robot i is selected as

$$\mathbf{q}^i_{\mathrm{p},k+1}(t) = [\mathbf{q}^i_{\mathrm{o},k+1}(t) \quad \mathbf{q}^i_{\mathrm{b},k+1}(t)],$$

while the trajectory for the robot j is either

$$\mathbf{q}^j_{\mathrm{p},k+1}(t) = [\mathbf{q}^j_{\mathrm{o},k+1}(t) \quad \mathbf{q}^j_{\mathrm{b},k+1}(t)] \quad \text{or} \quad \mathbf{q}^j_{\mathrm{p},k+1}(t) = [\mathbf{q}^j_{\mathrm{b},k}(t) \quad \mathbf{q}^j_{\mathrm{b},k+1}(t)].$$

In both cases, the continuous collision detection algorithm has been applied, which obviously guarantees that the condition

$$C^{i,j}(k + 1) = O(\mathbf{q}^i_{\mathrm{p},k+1}(t)) \cap O(\mathbf{q}^j_{\mathrm{p},k+1}(t)) = \varnothing, \quad \forall i \neq j, \tag{5.17}$$

holds.

Subcase 2.2 or Case 3:

The braking trajectories are respectively chosen for the robots i and j as

$$\mathbf{q}_{\mathrm{p},k+1}^{i}(t) = [\mathbf{q}_{\mathrm{b},k}^{i}(t) \quad \mathbf{q}_{\mathrm{b},k+1}^{i}(t)] \quad \text{and} \quad \mathbf{q}_{\mathrm{p},k+1}^{j}(t) = [\mathbf{q}_{\mathrm{b},k}^{j}(t) \quad \mathbf{q}_{\mathrm{b},k+1}^{j}(t)].$$

Since the condition $C^{i,j}(k) = \varnothing$ holds, one has

$$O(\mathbf{q}_{\mathrm{b},k}^{i}(t)) \cap O(\mathbf{q}_{\mathrm{b},k}^{j}(t)) = \varnothing. \tag{5.18}$$

Moreover, one has that $\mathbf{q}_{\mathrm{b},k+1}^{i}(t) \equiv \mathbf{q}_{\mathrm{b},k}^{i}(t_{\mathrm{c},k})$ and $\mathbf{q}_{\mathrm{b},k+1}^{j}(t) \equiv \mathbf{q}_{\mathrm{b},k}^{j}(t_{\mathrm{c},k})$ due to the standstill trajectory segments, which leads to

$$O(\mathbf{q}_{\mathrm{b},k+1}^{i}(t)) \cap O(\mathbf{q}_{\mathrm{b},k+1}^{j}(t)) = \varnothing. \tag{5.19}$$

From (5.18) and (5.19), one has that the condition

$$C^{i,j}(k+1) = O(\mathbf{q}_{\mathrm{p},k+1}^{i}(t)) \cap O(\mathbf{q}_{\mathrm{p},k+1}^{j}(t)) = \varnothing, \quad \forall i \neq j, \tag{5.20}$$

holds.

From (5.17) and (5.20), one has that

$$C^{i,j}(k+1) = O(\mathbf{q}_{\mathrm{p},k+1}^{i}(t)) \cap O(\mathbf{q}_{\mathrm{p},k+1}^{j}(t)) = \varnothing, \quad \forall i \neq j.$$

Therefore, the condition

$$C^{i,j}(k) = O(\mathbf{q}_{\mathrm{p},k}^{i}(t)) \cap O(\mathbf{q}_{\mathrm{p},k}^{j}(t)) = \varnothing, \quad \forall i \neq j,$$

holds for every planning cycle $k \geq 0$ by the mean of the induction principle. Since only the first segment $\mathbf{q}_{\mathrm{e},k}^{i,j}(t)$ of the trajectory concatenation $\mathbf{q}_{\mathrm{p},k}^{i,j}(t) = [\mathbf{q}_{\mathrm{e},k}^{i,j}(t) \quad \mathbf{q}_{\mathrm{b},k}^{i,j}(t)]$ is executed, the collision-free condition

$$S(t) = O(\mathbf{q}_{\mathrm{e}}^{i}(t)) \cap O(\mathbf{q}_{\mathrm{e}}^{j}(t)) = \varnothing, \quad \forall i, j \in \{1, ..., N_{\mathrm{r}}\}, \quad i \neq j, \quad \forall t,$$

is consequently always true ■.

If the collision avoidance algorithm starts at a *safe state*, e.g. all robots are standstill, then the initial condition (5.15) will obviously be satisfied. By incorporating the *back-up* braking trajectory segments, imminent collision states can be simply avoided and the invariant safety can always be guaranteed. Additionally, these braking trajectory segments can also be applied to stop the whole system in emergency cases.

5.6 A Conflict Resolution Algorithm

The invariant safety can be ensured by the collision avoidance algorithm. Nevertheless, there is no guarantee that each robot can reach its goal location. As illustrated in Figure 5.5 (left), if the robot 1 stays at its initial position, generating a feasible trajectory for the robot 2 will be impossible. Another example is depicted in Figure 5.5 (right). If the robot 1 trajectory is generated regardless of the robot 2, then the robot 1 will obstruct the robot 2 from reaching its goal. In both cases, desired optimal cooperative paths for the robots are respectively depicted by the dashed green and red lines. Therefore, a conflict resolution algorithm is employed in order to enable trajectory generation as well as optimize overall system performance. As mentioned in Chapter 2, the conflict resolution algorithm and the prioritized planning technique are based on a similar idea [6, 13, 25]. In fact, the conflict resolution algorithm is composed of two ingredients, which are an A^* search algorithm in approximated time-space configurations and an optimal planning sequence finding algorithm. In the following, these algorithms are explained.

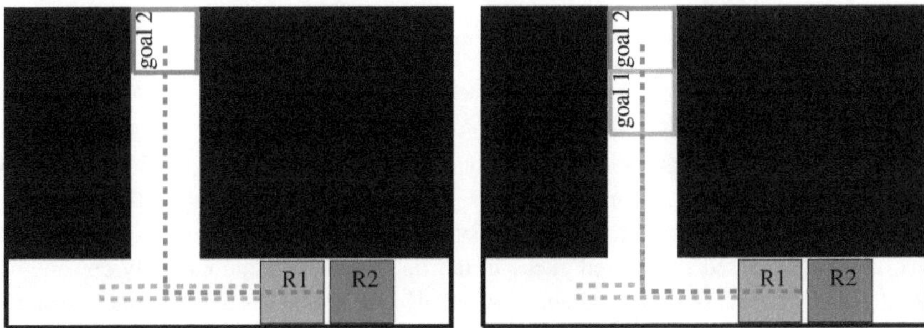

Figure 5.5: Application examples of the conflict resolution algorithm

5.6.1 A* Search Algorithm in Approximated Time-space Configurations

The time-space configuration has been initially developed for robot motion planning in an environment with dynamical obstacles, whose trajectories are known [25, 27]. By considering other preplanned robots as dynamical obstacles, the time-space configuration can be applied to resolve conflicts among robots [6, 13, 107]. Particularly, an approximate three dimensional grid-map is constructed, which has a time dimension and two x and y space dimensions. Moreover, the x and y space dimensions are exactly discretized by the robot length L, while the time dimension is sampled by $\Delta T = \frac{L}{v_{\max}}$, where $v_{\max} = v_{\max,x} = v_{\max,y}$ is the maximal velocity. Since other kinematic constraints $\mathbf{a}_{\max}$ and $\mathbf{j}_{\max}$ are not considered, this time-space configuration is only an approximation.

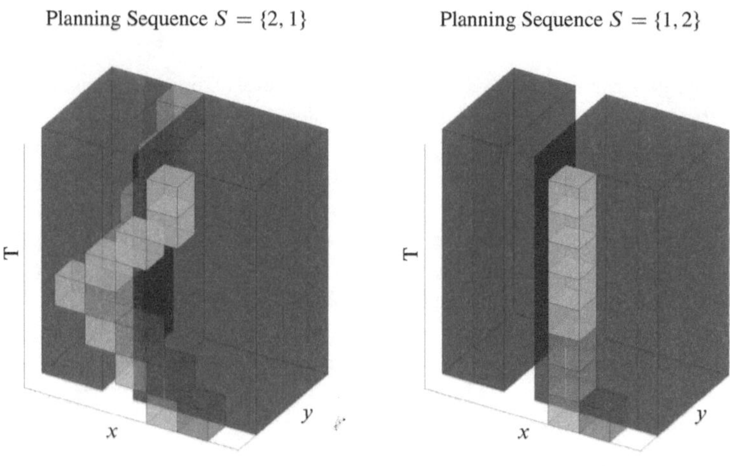

Figure 5.6: A* search algorithm in the approximated time-space configuration

In order to illustrate the time-space configuration, the above example in Figure 5.5 (left) is used. As depicted in Figure 5.6, the stationary obstacles are represented by the black cubes, which appear as parallel columns to the time axis. If the robot 2 trajectory is planned, the red path is obtained, which can be presented by the red cubes in the time-space configuration. By considering the robot 2 as a dynamical obstacle and applying A* algorithm, one has a three dimensional path of the robot 1 in the time-space configuration as shown in Figure 5.6 (left). This three dimensional path is represented by the green cubes in the time-space configuration. Eventually, the robot 1 optimal trajectory is obtained by projecting the three dimensional path in the x-y plane. In order to apply A* search algorithm in the time-space configuration, the time dimension must be distinctly considered. Since the time can only be increased, there are maximal nine candidate successors for each node (x, y, T) in A* search algorithm. These candidate successors are respectively $\big(x \pm \{0, L\}, y \pm \{0, L\}, T + \Delta T\big)$. Conversely, if the robot 1 path is firstly planned, a time-space configuration can be depicted in Figure 5.6 (right). Consequently, A* algorithm is not able to find a path for the robot 2. In other words, the planning sequence $S = \{1, 2\}$ is infeasible. Evidently, the planning sequence plays a decisive role in the feasibility and optimality. Therefore, an optimal planning sequence finding algorithm is subsequently introduced in the following.

5.6.2 The Optimal Planning Sequence Finding Algorithm

For a set of N_r robots, $N_r!$ possible planning sequences exist. In other words, the number of possible planning sequences increases more than exponentially with the number of robots. For example, if $N_r = 20$, the number of the possible planning sequences is approximately 2.4×10^{18}. Consequently, examining all planning sequences is impossible. In order to improve the search efficiency, the random-restart hill climbing algorithm and additional heuristic constraints are employed. These techniques were originally introduced by Maren Benewitz and Jur P. van den Berg [5, 6, 102], which will be revisited in the following.

The random-restart hill climbing algorithm has been successfully deployed in large or infinite search spaces [72, 84], whose working principle can be summarized in the following. Particularly, the algorithm is triggered by two variables, which are respectively *max_tries* and *max_flips*. The variable *max_tries* defines the number of iterations, which generates a new random planning sequence S_r. In each iteration, the new planning sequence is additionally modified *max_flips* times. In fact, two random integers x and y are created such as $1 \leq x < y \leq N_r$. Subsequently, the planning sequence is swapped at the positions x and y. From an original planning sequence

$$S_r = \{r_1, ..., r_x, ..., r_y, ..., r_{N_r}\},$$

one obtains a new modified planning sequence

$$S_{m,r} = \{r_1, ..., r_y, ..., r_x, ..., r_{N_r}\}.$$

After validating *max_tries* $\times$ *max_flips* planning sequences, the best solution S^* can be chosen, which results in the minimal overall travel time. In other words, our ultimate objective is optimizing the travel time, which is needed by the last robot in order to arrive its goal. Moreover, the search efficiency can be increased by incorporating the following additional heuristic constraints into the random-restart hill climbing algorithm.

In general, imposing the following heuristic constraints does not guarantee that the obtained planning sequences are feasible. However, the planning sequences satisfying the constraints come up with a feasible solution more often than those violating the constraints. Consequently, incorporating additional constraints can boost the search speed as well as increase the possibility of finding a feasible solution. These heuristic constraints are intuitively established, which nevertheless have been successfully deployed in several experiments [6, 102]. The first heuristic constraint is simple. Since the longest travel time from a start to a goal is minimized, the robot requiring the largest travel time without considering other robots should move unobstructedly [102]. In other words, this robot should be firstly planned due to the fact that other robots with expected shorter travel time can afford to avoid this robot.

Although the first heuristic contraint is effective, it fails to focus the search for a large number of robots operating in a confined working space. Therefore, a more powerful additional heuristic constraint is introduced by Maren Bennewitz [6], which can be stated as follows: *If the goal of a robot i blocks the shortest path of a robot j, then the robot j should be planned first.* As shown in Figure 5.7, the robot 1 blocks the shortest path of the robot 2, the robot 2 path should be planned

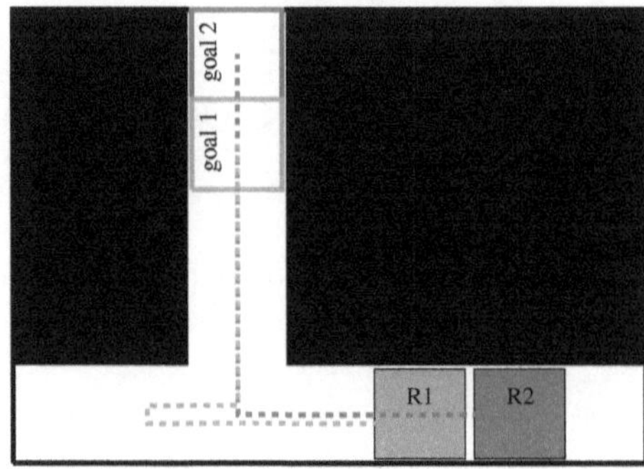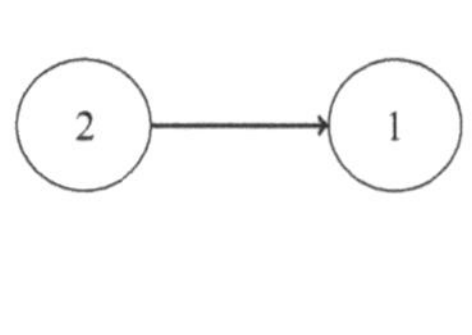

Figure 5.7: The second heuristic constraint and a planning sequence graph

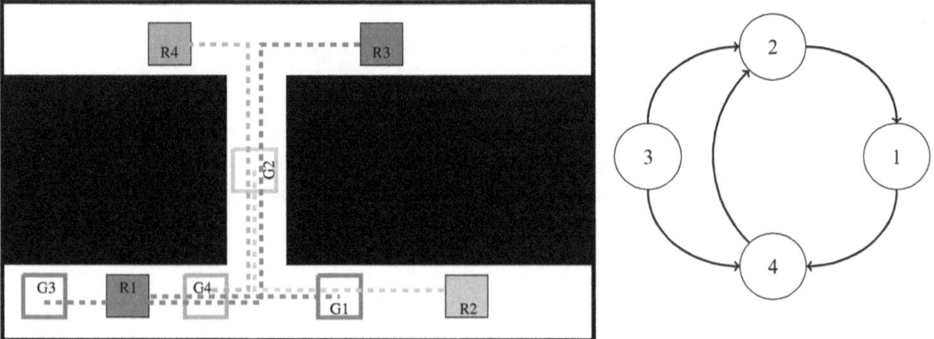

Figure 5.8: An example for cycle dependency. The set of robots is divided into two groups, which are $S_1 = \{3\}$ and $S_2 = \{1, 2, 4\}$. The group S_2 forms a cycle of dependency.

first according to this heuristic constraint. By applying this heuristic constraint, one obtains a planning sequence graph. In the planning sequence graph, if the robot i is planned before the robot j, a directed edge is constructed as shown in Figure 5.7.

Nevertheless, a cyclic dependency can exist, which implies that no planning sequence satisfies the heuristic constraint. For the sake of clarification, an example is depicted in Figure 5.8. In this case, the robots 1, 2 and 4 form a cycle dependency. Indeed, the goal 1 lies on the shortest path of the robot 2 and the goal 4 lies on the shortest path of the robot 1 while the goal 2 lies on the shortest path of the robot 4. Consequently, the set of robots can be divided into two groups S_1 and S_2. The first group S_1 contains all robots, which are not in the cycle. Furthermore, their path

---

Algorithm 5.2 Finding the optimal planning sequence S^*

    Input: The control variables *max_tries*, *max_flips* and the number k of iterations in the first search period.

    Output: The best found planning sequence S^*

1:   $S^* \leftarrow$ null;
2:   paths $\leftarrow$ null;
3:   $[S_1, S_2] =$ setConstraints(S);   *// split the robots into two groups by using the heuristics*
4:   count = 1;   *// a variable used to divide the search periods*
5:   **for** $i = 1 : 1 : max\_Tries$ **do**
6:      **if** (count $\leq k$) **then**
7:         $S_{2,m,r}$ generateRandomSequence(S_2);   *// in the first search period*
8:         $S_{m,r}$ $[S_1 \quad S_{2,m,r}]$;
9:      **else**
10:        $S_{m,r}$ generateRandomSequence(S);   *// in the second search period*
11:      **end if**
12:      **for** $j = 1 : 1 : max\_flips$ **do**
13:        **if** (count $\leq k$) **then**
14:          $\tilde{S}_{2,m,r}$ flip($S_{2,m,r}$);
15:          $\tilde{S}_{m,r}$ $[S_1 \quad \tilde{S}_{2,m,r}]$;
16:        **else**
17:          $\tilde{S}_{m,r} \leftarrow$ flip($S_{m,r}$);
18:        **end if**
19:        newpaths   computeConflictFreePaths($\tilde{S}_{m,r}$);
20:        **if** (paths $==$ null) $\vee$ (travelTime(paths) $>$ travelTime(newPaths)) **then**
21:          paths   newpaths;
22:          S^* $\tilde{S}_{m,r}$;
23:        **end if**
24:        count $\leftarrow$ count $+ 1$;
25:      **end for**
26: **end for**

---

should be planned before all robots in the dependency cycle. The second group S_2 contains the remaining robots. In order to obtain an optimal planning sequence, the random-restart hill-climbing algorithm can be applied for the group S_2, which has a smaller number of robots than the original group S. Therefore, the search speed and efficiency can be accelerated.

Obviously, a better solution can be obtained without constraining the search space in the group S_2. However, constraining the search space in S_2 increases the possibility of finding a feasible and better solution at the beginning of the search procedure [6]. Therefore, the search procedure is divided into two periods. Initially, the random-restart hill-climbing algorithm is applied for the robot group S_2. Afterwards, the original algorithm is employed. Eventually, the pseudocode of the search procedure is shown in Algorithm 5.2.

5.7 Algorithms Simulation

In this section, these algorithms are validated in MATLAB® 2017b environments. Firstly, the invariant safety of the centralized collision avoidance algorithm is verified. A simple simulator illustrated in Figure 5.9 is created, which has a length of 4 m, a width of 2.5 m and four identical square robots. Particularly, each robot has a width of 0.12 m. These robots are assigned to translate from starts to goals and back again. The starts and goals are randomly permutated during the simulation. The robot kinematic constraints are identical, which are respectively $\mathbf{v}_{max} = \begin{bmatrix} 3 & 3 \end{bmatrix}^T \frac{m}{s}$, $\mathbf{a}_{max} = \begin{bmatrix} 20 & 20 \end{bmatrix}^T \frac{m}{s^2}$ and $\mathbf{j}_{max} = \begin{bmatrix} 800 & 800 \end{bmatrix}^T \frac{m}{s^3}$. The sampling time is $T_s = 5ms$ while the planning cycle of the centralized collision avoidance algorithm is $T_p = 70ms = 14T_s$. The priorities of all robots are fixed and the whole simulation took about 12s.

Time = 0.695 s

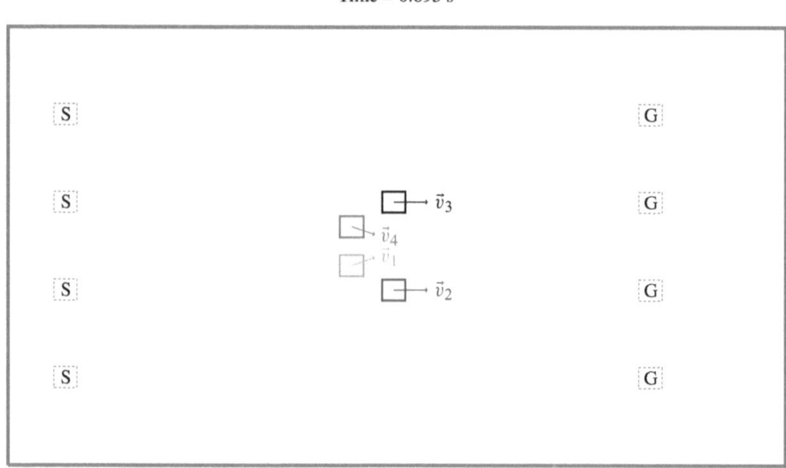

Figure 5.9: The environment for verifying the centralized collision avoidance algorithm

As expected, no collision is found during the simulation. Furthermore, all kinematic constraints are satisfied. The simulation confirmed the invariant safety guarantee. Additionally, scalability of the centralized collision avoidance algorithm is also examined in another experiment. In order to simulate the worst case, all robots are commisioned to translate in a grid pattern. The numbers of robots range from 2 to 10. All measurements were taken on a PC with a Intel® Core™ i3 − 540 CPU 3.06 GHz. In Figure 5.10, the result is depicted. On average, the centralized collision avoidance algorithm requires only 4.1 ms for two robots and 144.4 ms for ten robots. In this experiment, the planning cycle T_p is set to be 200 ms. As expected, the computing time of the centralized collision avoidance algorithm increases quadratically with the number of robots. According to the best of our knowledge, there is no any existing motion planning algorithm, which is able to be

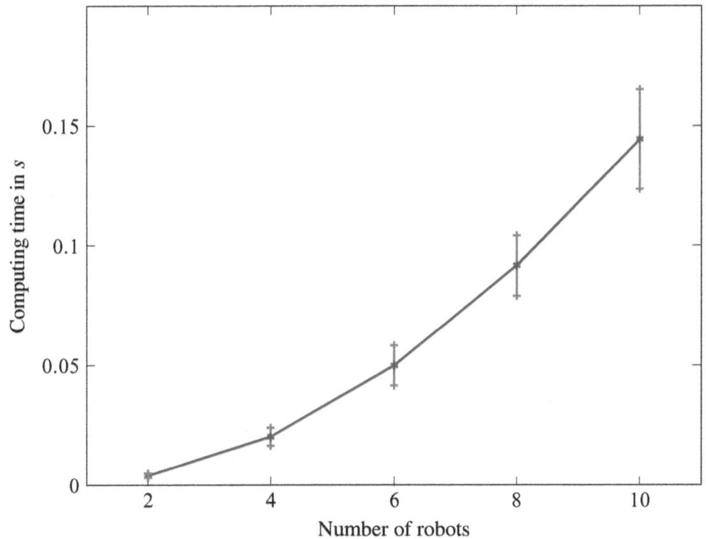

Figure 5.10: The minimal, maximal and average computing time of the centralized collision avoidance algorithm

executed with such a small computing burden and simultaneously considers the high-order time derivatives constraints such as jerks. Although the centralized collision avoidance algorithm was not compiled in C, it can still be executed within a planning cycle T_p.

Finally, the conflict resolution algorithm is verified. First, the algorithm is used to solve the planning problem depicted in Figure 5.8. The planning sequence $S = \{3, 4, 1, 2\}$ is obtained, which belongs to the search space constrained by the aforementioned heuristics. Particularly, this planning sequence is the only feasible solution for the problem. Consequently, the efficiency of employing the heuristics is ascertained. Moreover, the algorithms are applied to solve more complex scenarios with much more number of robots. Videos demonstrating motions of the centralized coordination solution is attached and can be found online. [1,2]

5.8 Conclusion

In this chapter, a highly modular novel centralized coordination solution is presented, which is comprised of a collision avoidance algorithm and a conflict resolution algorithm. The invariant safety is guaranteed by the collision avoidance algorithm, while deadlocks can be resolved by the conflict resolution algorithm. Moreover, the effectiveness of the centralized coordination solution is also confirmed by several simulations.

In the next step, we want to apply the developed algorithms to real world systems. Without any further effort, the centralized collision avoidance algorithm can be directly integrated into commercial high-order smooth trajectory motion control systems [8]. For the real world deployment, an extra small margin is added to robots due to tracking errors of motion controllers. Since the robot heterogeneous kinematic constraints is not considered by the conflict resolution algorithm, an automated finite-state machine generation is required in order to accomplish the full solution.

Nevertheless, the centralized coordination solution can only be applied to a moderate number of robots due to its polynomial computational complexity. Therefore, there is a substantial motivation to develop a distributed coordination solution. In order to scale the system to another dimension, a novel distributed collision avoidance algorithm is presented in the next chapter.

6 An Asynchronous Distributed Collision Avoidance Algorithm

This chapter presents a novel asynchronous distributed collision avoidance algorithm for a group of autonomous passive robots translating on a grid of rectangular intelligent agents. The algorithm is not only able to guarantee collision-free motions but also robust with respect to stochastic arbitrarily large communication time delays among agents. Consequently, one is able to deploy the algorithm to the aforementioned distributed system architecture. Finally, efficiency and applicability of the algorithm are demonstrated by simulations.

6.1 Introduction

As aforementioned in Chapter 1, several industrial material handling systems can be modeled by the distributed system architecture. As illustrated in Figure 6.1, the whole system consists of numerous passive robots, which translate on a grid of intelligent agents. In contrast to the centralized system architecture, the agent and industrial PC clocks can not be synchronized. Moreover, communication time delays can be stochastic and arbitrarily large. In this architecture, the robot ultimate goal locations are assigned by the industrial PC, while the trajectory generation algorithm

Industrial PC

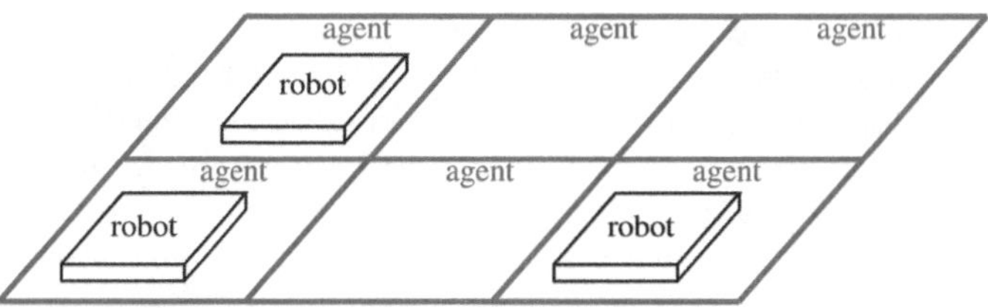

Figure 6.1: The distributed system architecture

as well as the distributed coordination solution are executed by the agents. In other words, each agent is responsible for planning trajectories of robots, which are placed on the agent.

In analogy to the centralized one, the distributed coordination solution should operate as an additional process to the trajectory optimization presented in Chapter 4. As mentioned in Chapter 2, a collision avoidance algorithm is the key ingredient to a distributed coordination solution. In the following, an asynchronous distributed collision avoidance algorithm is presented in detail. Eventually, the chapter is structured by the following sections:

- **Preliminaries and Nomenclatures:** All preliminaries and nomenclatures for describing a robotic agent network and designing an asynchronous distributed collision avoidance algorithm are provided.

- **A Fundamental Working Principle:** The fundamental working principle of the distributed collision avoidance algorithm is explained.

- **An Information Update Scheme:** The information update scheme is introduced in this section, which is used to handle stochastic communication delays.

- **The Distributed Collision Avoidance Algorithm Description:** All basic functions of the distributed collision avoidance algorithm are described in detail.

- **A Formal Safety Guarantee Proof:** The invariant safety guarantee of the distributed collision avoidance algorithm is formally proven.

- **Computational Complexity and Simulations:** A computational complexity analysis of the distributed collision avoidance algorithm is given. Moreover, applicability of the distributed collision avoidance algorithm is demonstrated by simulations.

- **Conclusion:** Finally, the results are concluded and future works are mentioned in order to accomplish the distributed coordination solution.

6.2 Preliminaries and Nomenclatures

In this section, all preliminaries as well as nomenclatures are provided, which are used for describing the robotic agent network and designing the asynchronous distributed collision avoidance algorithm in the subsequent passages. Particularly, the distributed collision avoidance algorithm is derived based on a standard robotic network framework [14, 65, 66]. As shown in Figure 6.2, the distributed collision avoidance algorithm is composed of a state transition function (stf), a trajectory generation function (tgf) and a message generation function (msf), which will be briefly explained in the following.

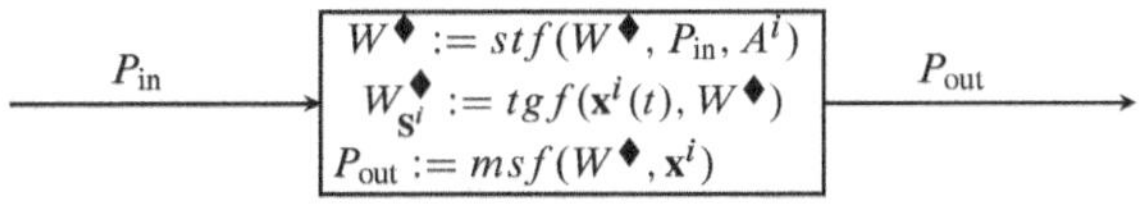

$$W^{\blacklozenge} := stf(W^{\blacklozenge}, P_{\text{in}}, A^i)$$
$$W_{\mathbf{s}^i}^{\blacklozenge} := tgf(\mathbf{x}^i(t), W^{\blacklozenge})$$
$$P_{\text{out}} := msf(W^{\blacklozenge}, \mathbf{x}^i)$$

$P_{\text{in}} \longrightarrow \qquad \longrightarrow P_{\text{out}}$

Figure 6.2: The distributed collision avoidance algorithm

Let denote $W^{\blacklozenge}$ as the state variables of the agent $\blacklozenge$. These variables are given in Table 6.1, whose usages will be precisely explained in the next sections. In order to ensure the invariant safety, the agents are required to exchange information compressed in messages, whose elements are provided in Table 6.2. These messages P_{out} are generated and sent by the message generation function (msf). Additionally, an goal is assigned to a robot i via another message A^i given in Table 6.3 from the industrial computer. Hinging upon the receiving messages P_{in} and A^i, each agent $\blacklozenge$ updates its state variables $W^{\blacklozenge}$ by the state transition function (stf). Eventually, the trajectory generation function (tgf) can be applied for the robot i placed on the agent $\blacklozenge$ based on the local state variables $W^{\blacklozenge}$ and the robot state $\mathbf{x}^i(t)$.

Table 6.1: A list of state variables $W^{\blacklozenge}$ of an agent $\blacklozenge$

Variable	Description
i, j	are the robot identification numbers.
$\mathbf{q}^i(t), \dot{\mathbf{q}}^i(t), \ddot{\mathbf{q}}^i(t)$	is the robot i position, velocity and acceleration.
$\mathbf{x}^i(t)$	is the robot i state, i.e $\mathbf{x}^i(t) = \begin{bmatrix} \mathbf{q}^i(t) & \dot{\mathbf{q}}^i(t) & \ddot{\mathbf{q}}^i(t) \end{bmatrix}^T$.
$W_{\mathcal{M}}^{\blacklozenge}$	is the set of robots concerned with the agent $\blacklozenge$.
$W_{\mathcal{M}_{\text{on}}}^{\blacklozenge}$	is the set of robots placed on the agent $\blacklozenge$, which implies $W_{\mathcal{M}_{\text{on}}}^{\blacklozenge} \subseteq W_{\mathcal{M}}^{\blacklozenge}$.
$W_{\mathbf{A}_i}^{\blacklozenge}$	is the robot i total reserved area on the working environment.
$W_{\mathbf{B}_i}^{\blacklozenge}$	is the robot i total requested area on the working environment.
$W_{\mathbf{A}_{i,\blacklozenge}}^{\blacklozenge}$	is the robot i reserved area on the agent $\blacklozenge$.
$W_{\mathbf{B}_{i,\blacklozenge}}^{\blacklozenge}$	is the robot i requested area on the agent $\blacklozenge$.
$W_{\text{area}}^{\blacklozenge}$	is the area of the agent $\blacklozenge$.
$W_{\mathbf{q}_{\text{goal}}^i}^{\blacklozenge}$	is the robot i goal location, $\forall i \in W_{\mathcal{M}_{\text{on}}}^{\blacklozenge}$.
$x_{\text{global}}^{\blacklozenge}$	is the global x-coordinate of the agent $\blacklozenge$.
$y_{\text{global}}^{\blacklozenge}$	is the global y-coordinate of the agent $\blacklozenge$.
$W_{\mathbf{s}^i}^{\blacklozenge}$	is the robot i trajectory segment, $\forall i \in W_{\mathcal{M}_{\text{on}}}^{\blacklozenge}$.
$W_{t_{\text{last},\clubsuit}}^{\blacklozenge}$	is the last time stamp according to the agent $\clubsuit$ local clock of a message, which was sent by the agent $\clubsuit$ and received by the agent $\blacklozenge$.
$W_{\text{counter},i}^{\blacklozenge}$	is the robot i requested update counter, $\forall i \in W_{\mathcal{M}}^{\blacklozenge}$.
$W_{\text{cost},i}^{\blacklozenge}$	is the robot i cost with respect to the agent $\blacklozenge$, $\forall i \in W_{\mathcal{M}}^{\blacklozenge}$.
$W_{t_{\text{local,clock}}}^{\blacklozenge}$	is the time according to the agent $\blacklozenge$ local clock.

Table 6.2: A list of variables in a message $\mathbf{P}_{\{\mathrm{in,\ out}\}}$

Variable	Description
$\mathbf{P}^{\mathrm{id}}_{\{\mathrm{in,\ out}\}}$	is the sending agent identification number.
$\mathbf{P}_{\{\mathrm{in,\ out}\},t_{\mathrm{last}}}$	is the time stamp of the sending agent clock.
$\mathbf{P}_{\{\mathrm{in,\ out}\},i}$	is the robot identification number.
$\mathbf{P}_{\{\mathrm{in,\ out}\},\mathbf{x}^i}$	is the robot i state.
$\mathbf{P}_{\{\mathrm{in,\ out}\},\mathrm{counter},i}$	is the robot i requested update counter.
$\mathbf{P}_{\{\mathrm{in,\ out}\},\mathbf{B}_i}$	is the robot i requested area on the working environment.
$\mathbf{P}_{\{\mathrm{in,\ out}\},\mathbf{A}_i,\blacklozenge}$	is the robot i reserved area on the sending agent $\blacklozenge$.

Table 6.3: A set of variables in an assignment message A

Variable	Description
A^i	the robot i identification number.
$A_{\mathbf{q}^i_{\mathrm{goal}}}$	is the robot i goal location.

In an asynchronous robotic network, communication time delay among agents can be stochastic and arbitrarily large. Additionally, the agent local clocks are not synchronized as aforementioned. Therefore, the distributed collision avoidance algorithm is executed asynchronously by the agents. In consequence, it is extremely challenging to develop a distributed collision avoidance algorithm. Nevertheless, collision-free trajectories can still be generated by elegantly designing communication protocols and a mechanism, which will be introduced in the following sections.

6.3   A Fundamental Working Principle

A reasonable idea is that each agent $\blacklozenge$ only plans trajectories for all robots, whose central points are located on the agent area $W^{\blacklozenge}_{\mathrm{area}}$. Based on the total reserved area $W^{\blacklozenge}_{\mathbf{A}_i}$ on the working environment, trajectories are generated for the robot i such that the occupied area are always inside the reserved area $W^{\blacklozenge}_{\mathbf{A}_i}$. Consequently, collision-free motions can be guaranteed by ensuring the reserved area distinction between any two robots as

$$W^{\blacklozenge}_{\mathbf{A}_i}(t) \cap W^{\blacklozenge}_{\mathbf{A}_j}(t) = \varnothing, \quad \forall t, \quad \forall i \neq j, \quad i \in W^{\blacklozenge}_{\mathcal{M}_{\mathrm{on}}}, \quad j \in W^{\blacklozenge}_{\mathcal{M}_{\mathrm{on}}}.$$

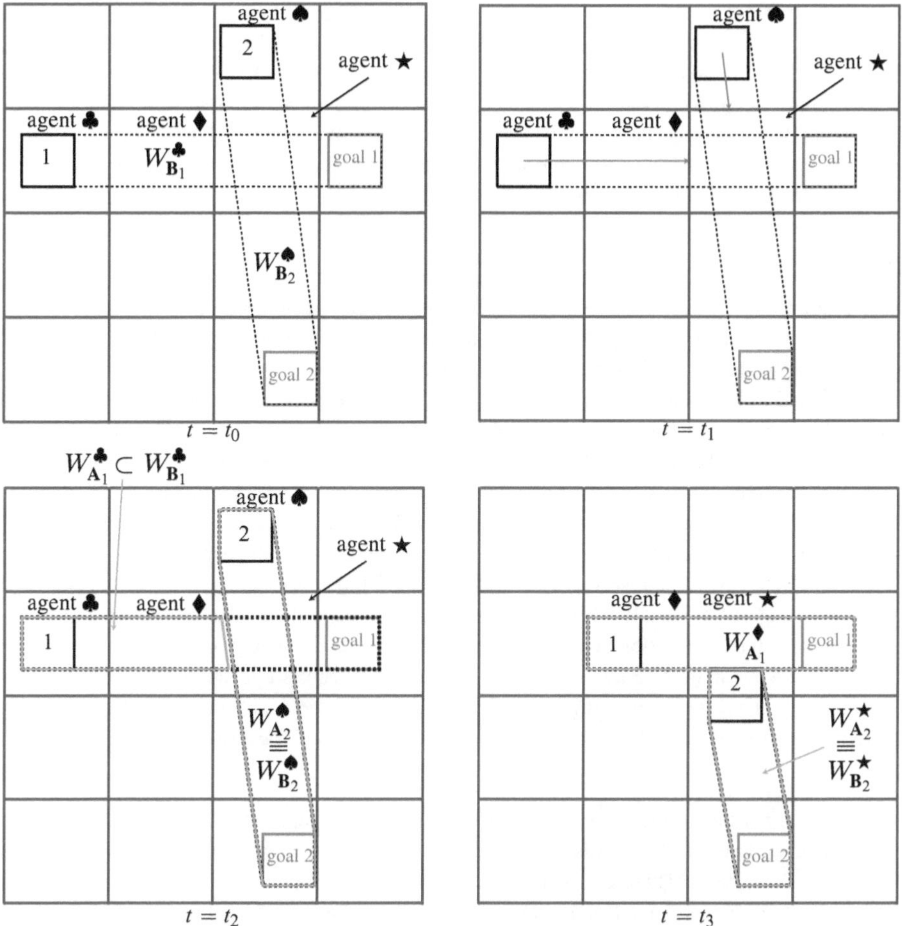

Figure 6.3: The working principle of the distributed collision avoidance algorithm

Therefore, assigning the reserved areas is a key to the distributed collision avoidance algorithm. In the following, an assigning procedure is explained in detail.

Basically, the reserved areas assigning procedure is similar to Oliver Purwin's idea [77, 78]. In lieu of applying to mobile robots, the procedure is modified for the static agents. Instead of computing the reserved area $W_{\mathbf{A}_i}^{\spadesuit}$ directly, agent ♦ indicates a requested area $W_{\mathbf{B}_i}^{\spadesuit}$ for a robot $i \in W_{\mathcal{M}_{\mathrm{on}}}^{\spadesuit}$ based on its initial state $\mathbf{x}^i(t)$ as well as a goal location $\mathbf{q}_{\mathrm{goal}}^i$. Subsequently, the requested area $W_{\mathbf{B}_i}^{\spadesuit}$ is compressed in a message as $P^{\blacklozenge}_{\mathrm{out},\mathbf{B}_i}$ and sent to other related agents ♠. After the related agents ♠ receive the messages confirming that $P^{\spadesuit}_{\mathrm{out},\mathbf{B}_i} \cap W^{\spadesuit}_{\mathrm{area}} \neq \varnothing$, they will eventually compute the local reserved area $W^{\spadesuit}_{\mathbf{A}_i,\spadesuit}$ with $W^{\spadesuit}_{\mathbf{A}_i,\spadesuit} \subseteq W^{\spadesuit}_{\mathrm{area}}$ for the robot i. Moreover, the local reserved area $W^{\spadesuit}_{\mathbf{A}_i,\spadesuit}$ is delivered back to an agent ♣, on which the robot i is currently located. Note that the agent ♣ is not necessarily identical to the original agent ♦ due to communication time delays and the robot i movements. Eventually, the agent ♣ will update the reserved area $W^{\clubsuit}_{\mathbf{A}_i}$ for the robot i based on its state variables and the receiving information. In the following, an example is provided to demonstrate the assigning procedure.

As shown in Figure 6.3, agents ♣ and ♠ receive assignment messages from the industrial computer at a time instant t_0, which requires both robots 1 and 2 to translate to new goal locations. Subsequently, the agents ♣ and ♠ indicate the intentions by updating requested areas $W^{\clubsuit}_{\mathbf{B}_1}$, $W^{\spadesuit}_{\mathbf{B}_2}$ and inform other related agents via sending messages. Assume that an agent ★ receives both messages $P^{\clubsuit}$ and $P^{\spadesuit}$ at a time instant t_1. Then, the agent ★ will decide that the robot 2 has a higher priority by simply comparing distances from the robot positions in $\mathbf{P}^{\clubsuit}_{\mathrm{x}1}$ and $\mathbf{P}^{\spadesuit}_{\mathrm{x}2}$ to the agent ★. Consequently, the robot 2 has its desired reserved area on the agent ★. Contrarily, the robot 1 has its desired reserved area except the robot 2 reserved area. These reserved areas are sent back to the agents ♣ and ♠. At a time instant t_2, the agents ♣ and ♠ will update the reserved area $W^{\clubsuit}_{\mathbf{A}_1}$ and $W^{\spadesuit}_{\mathbf{A}_2}$ based on the feedback messages. These new reserved areas are depicted by green lines in Figure 6.3. Based on these reserved areas, robot 1 and 2 trajectories are generated. Furthermore, these areas $W^{\clubsuit}_{\mathbf{B}_1}$, $W^{\clubsuit}_{\mathbf{A}_1}$ $W^{\spadesuit}_{\mathbf{B}_2}$, $W^{\spadesuit}_{\mathbf{A}_2}$ shrink, when the robots follow their trajectories. As long as the robot 2 passes the intersection area, the agent ★ will assign a new reserved area for the robot 1 and inform an agent ♦. At a time instant t_3, the agent ♦ receives a message from the agent ★. Consequently, the reserved area $W^{\blacklozenge}_{\mathbf{A}_1}$ will be extended toward the goal 1 as shown in Figure 6.3 at the time instant t_3. Finally, the robots 1 and 2 can reach their goal locations.

Nevertheless, stochastic communication time delays among agents can deteriorate the proposed assigning procedure. In other words, receiving messages are not reliable due to the stochastic communication time delays. Therefore, a message $\mathbf{P}$ is only accepted if certain conditions are satisfied. These conditions are explained by an information update scheme in the following section.

6.4   An Information Update Scheme

In order to ensure that a message is relevant, a communication model is required and described in the following. In Figure 6.4, a communication cycle for a robot i is depicted. Basically, time instants t_0, t_1, t_2 and t_3 are denoted as milestones in the communication cycle. Let assume that a new area for the robot i placed on an agent ♦ is requested at the time instant t_0, i.e. $W^{\blacklozenge}_{\mathbf{B}_i,\mathrm{new}} \nsubseteq W^{\blacklozenge}_{\mathbf{B}_i}$. Consequently, the robot i update counter is increased as $W^{\blacklozenge}_{\mathrm{counter},i} := W^{\blacklozenge}_{\mathrm{counter},i} + 1$ and the agent ♦ will send a message $P^{\blacklozenge}$ to another related agent ♣. At the time instant t_1, the message arrives at the related agent ♣. Correspondingly, the related agent ♣ reacts by sending back a message

$P^{\clubsuit}$ at the time instant t_2. Furthermore, the message $P^{\clubsuit}$ is received by an agent $\spadesuit$ at the time instant t_3, which accomplishes the communication cycle. Note that at the time instant t_3, the robot i can be placed on the agent $\spadesuit$, which is not necessarily identical to the initial agent $\blacklozenge$ due to communication time delays. Therefore, the message $P^{\clubsuit}$ from the agent $\clubsuit$ is only accepted by the agent $\spadesuit$, if the two following conditions are satisfied:

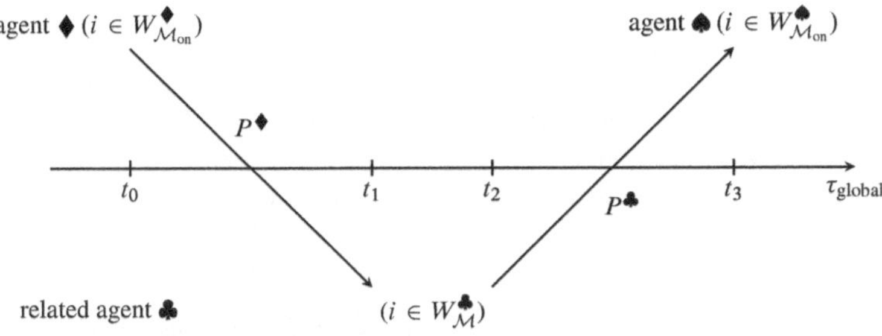

Figure 6.4: An information update scheme for a robot i

- Since the robot i can change its requested area during waiting for responses from other agents, the first condition is thereby to guarantee that the message contains a contemporary update for the robot i, i.e. $P^{\clubsuit}_{\mathrm{counter},i} \geq W^{\spadesuit}_{\mathrm{counter},i}$.

- Due to stochastic communication delays, messages from the agent $\clubsuit$ do not arrive at the agent $\spadesuit$ in a *first out first in* sequence. Consequently, the last time-stamp from the agent $\clubsuit$ according to its local clock is stored as $W^{\spadesuit}_{t_{\mathrm{last}},\clubsuit}$ in the agent $\spadesuit$. Therefore, a message $P^{\clubsuit}$ from the agent $\clubsuit$ to the agent $\spadesuit$ is accepted if $P^{\clubsuit}_{t_{\mathrm{last}}} > W^{\spadesuit}_{t_{\mathrm{last}},\clubsuit}$. As a result, the last time-stamp on the agent $\spadesuit$ is replaced by $W^{\spadesuit}_{t_{\mathrm{last}},\clubsuit} := P^{\clubsuit}_{t_{\mathrm{last}}}$. Conversely, $P^{\clubsuit}_{t_{\mathrm{last}}} \leq W^{\spadesuit}_{t_{\mathrm{last}},\clubsuit}$ implies that the message $P^{\clubsuit}$ is out of order. In other words, the agent $\spadesuit$ has received another message $\tilde{P}^{\clubsuit}$, which was sent by the agent $\clubsuit$ after $P^{\clubsuit}$.

Under the trivial assumption that communication time delays are greater than zero, irrelevant messages can be discarded by applying these two conditions. Based on the working principle and the information update scheme, the distributed collision avoidance algorithm can be designed and precisely described in the following.

6.5 The Distributed Collision Avoidance Algorithm Description

As aforementioned, the distributed collision avoidance algorithm is comprised of the state transition, trajectory generation and message generation functions, which are respectively explained in the following.

6.5.1 State Transition Function

The agent state variables are updated by the state transition function based on receiving messages from other agents and the industrial computer. As shown in Algorithm 6.1, the update procedure can be briefly described in the following.

If an agent $\blacklozenge$ receives a message $P^\clubsuit$ from another agent $\clubsuit$, it will firstly verify whether the message is valid. If the condition $P_{t_{\text{last}}}^\clubsuit > W_{t_{\text{last}},\clubsuit}^\blacklozenge$ is not fulfilled, the message $P^\clubsuit$ is out of order and consequently discarded according to the information update scheme. In other words, there is at least one message, which was sent by the agent $\clubsuit$ after $P^\clubsuit$ and received by the agent $\blacklozenge$. Contrarily, if the message $P^\clubsuit$ is accepted, the agent state variables will be proceeded according to the following cases:

Case 1: A robot $i = P_i^\clubsuit$ is an element of the robot set $W_\mathcal{M}^\blacklozenge$ concerned with the agent $\blacklozenge$, i.e. $P_i^\clubsuit \in W_\mathcal{M}^\blacklozenge$. Additionally, the condition $W_{\text{counter},i}^\blacklozenge \leq P_{\text{counter},i}^\clubsuit$ is satisfied, which guarantees that $P^\clubsuit$ is a contemporary update for the robot i. Moreover, the intersection between the robot i requested region $P_{\mathbf{B}_i}^\clubsuit$ and the agent $\blacklozenge$ area is not empty. According to lines 6-10 of Algorithm 6.1, the state variables for the robot i are updated by using information in the message. In fact, the new robot i reserved area $W_{\mathbf{A}_i,\blacklozenge}^\blacklozenge$ on the agent $\blacklozenge$ is replaced by the intersection of the previous reserved area $W_{\mathbf{A}_i,\blacklozenge}^\blacklozenge$ and the updated requested area $P_{\mathbf{B}_i}^\clubsuit$ as

$$W_{\mathbf{B}_i}^\blacklozenge := P_{\mathbf{B}_i}^\clubsuit \quad \text{and} \quad W_{\mathbf{A}_i,\blacklozenge}^\blacklozenge := W_{\mathbf{A}_i,\blacklozenge}^\blacklozenge \cap W_{\mathbf{B}_i}^\blacklozenge.$$

Case 2: In contrast to **Case 1**, the intersection between the robot i requested region $P_{\mathbf{B}_i}^\clubsuit$ and the agent area is empty, which implies that the robot i does not intend to pass the agent $\blacklozenge$ any more. As shown line 12 of Algorithm 6.1, the robot i state variables can consequently be removed from the agent.

Case 3: As stated in lines 16-20 of Algorithm 6.1, if the robot i is not an element of the robot set $W_\mathcal{M}^\blacklozenge$, i.e. $P_i^\clubsuit \notin W_\mathcal{M}^\blacklozenge$, then the new robot i state variables will be added to the agent.

Afterwards, the state variables are updated based on the message A^i from the industrial computer. If the robot i is currently placed on the agent, a new goal location is assigned. Subsequently, an area $\mathbf{B}_{i,\text{estimate}}$ is estimated, which encompasses the occupied area by the trajectory translating from the current robot state $\mathbf{x}^i(t)$ to the new goal location. As presented in Chapter 4, the trajectory optimization relies on the quintic Bézier splines, which guarantees that a generated trajectory always lies within a set of convex hulls defined by the control vertices. Therefore, the area $\mathbf{B}_{i,\text{estimate}}$ can be simply approximated by these convex hulls. Moreover, the robot i current reserved area $W_{\mathbf{A}_i}^\blacklozenge$ can not be arbitrarily reduced, since it can lead to an infeasible trajectory generation problem due to the kinematic constraints. Therefore, a candidate requested area $\mathbf{B}_{i,\text{candidate}}$ is rendered as a convex hull containing both $\mathbf{B}_{i,\text{estimate}}$ and $W_{\mathbf{A}_i}^{\blacklozenge}$. Subsequently, the local requested and reserved areas $W_{\mathbf{B}_i,\blacklozenge}^{\blacklozenge}$, $W_{\mathbf{A}_i,\blacklozenge}^{\blacklozenge}$ are updated according to lines 28-30. Moreover, if the new candidate requested area $\mathbf{B}_{i,\text{candidate}}$ is not a subset of the previous one, then the update counter variable $W_{\text{counter},i}^{\blacklozenge}$ is increased by one.

Algorithm 6.1 State transition function for an agent $\blacklozenge$

$\quad$ **Input:** agent state variables $W^{\blacklozenge}$, input messages $P^{\clubsuit}$, A^i from other agents and the industrial computer
$\quad$ **Output:** new agent state variables $W^{\blacklozenge}$
$\quad$ *Update state variables based on a received message $P^{\clubsuit}$ from another agent $\clubsuit$*

1: **if**$(P_{t_{\text{last}}}^{\clubsuit} > W_{t_{\text{last}},\clubsuit}^{\blacklozenge})$
2: $\quad\quad$ $W_{t_{\text{last}},\clubsuit}^{\blacklozenge} \leftarrow P_{t_{\text{last}}}^{\clubsuit}$;
3: $\quad\quad$ **if**$(P_i^{\clubsuit} \in W_{\mathcal{M}}^{\blacklozenge})$
4: $\quad\quad\quad$ **if**$(W_{\text{counter},i}^{\blacklozenge} \leq P_{\text{counter},i}^{\clubsuit})$
5: $\quad\quad\quad\quad$ **if**$(P_{\mathbf{B}_i}^{\clubsuit} \cap W_{\text{area}}^{\blacklozenge} \neq \varnothing)$
$\quad\quad\quad\quad\quad$ *Case 1: update the agent state variables*
6: $\quad\quad\quad\quad\quad$ $W_{\mathbf{B}_i}^{\blacklozenge} \leftarrow P_{\mathbf{B}_i}^{\clubsuit}$;
7: $\quad\quad\quad\quad\quad$ $W_{\mathbf{B}_i,\blacklozenge}^{\blacklozenge} \leftarrow P_{\mathbf{B}_i}^{\clubsuit} \cap W_{\text{area}}^{\blacklozenge}$;
8: $\quad\quad\quad\quad\quad$ $W_{\mathbf{A}_i,\blacklozenge}^{\blacklozenge} \leftarrow W_{\mathbf{B}_i,\blacklozenge}^{\blacklozenge} \cap W_{\mathbf{A}_i,\blacklozenge}^{\blacklozenge}$;
9: $\quad\quad\quad\quad\quad$ $W_{\text{counter},i}^{\blacklozenge} \leftarrow P_{\text{counter},i}^{\clubsuit}$;
10: $\quad\quad\quad\quad\quad$ $W_{\text{cost},i}^{\blacklozenge} \leftarrow \text{distance}(P_{\mathbf{x}^i}^{\clubsuit} \rightarrow W^{\blacklozenge})$;
11: $\quad\quad\quad\quad$ **else**
$\quad\quad\quad\quad\quad$ *Case 2: remove the robot i from the agent*
12: $\quad\quad\quad\quad\quad$ $W^{\blacklozenge} \leftarrow \text{removeRobotFromAgent}(P_i^{\clubsuit})$;
13: $\quad\quad\quad\quad$ **end if**
14: $\quad\quad\quad$ **end if**
15: $\quad\quad$ **else**
$\quad\quad\quad\quad\quad$ *Case 3: add the robot i to the agent*
16: $\quad\quad\quad$ $W_{\mathcal{M}}^{\blacklozenge} \leftarrow W_{\mathcal{M}}^{\blacklozenge} \cup \{P_i^{\clubsuit}\}$;
17: $\quad\quad\quad$ $W_{\mathbf{B}_i}^{\blacklozenge} \leftarrow P_{\mathbf{B}_i}^{\clubsuit}$;
18: $\quad\quad\quad$ $W_{\mathbf{B}_i,\blacklozenge}^{\blacklozenge} \leftarrow P_{\mathbf{B}_i}^{\clubsuit} \cap W_{\text{area}}^{\blacklozenge}$;
19: $\quad\quad\quad$ $W_{\text{counter},i}^{\blacklozenge} \leftarrow P_{\text{counter},i}^{\clubsuit}$;
20: $\quad\quad\quad$ $W_{\text{cost},i}^{\blacklozenge} \leftarrow \text{distance}(P_{\mathbf{x}^i}^{\clubsuit} \rightarrow W^{\blacklozenge})$;
21: $\quad\quad$ **end if**
22: **end if**
$\quad$ *Update state variables bases on a message A^i from the industrial computer*
23: **if**$(A^i \in W_{\mathcal{M}_{\text{on}}}^{\blacklozenge}) \wedge (\text{isValid}(A^i))$
24: $\quad\quad$ $W_{\mathbf{q}_{\text{goal}}^i}^{\blacklozenge} \leftarrow A_{\mathbf{q}_{\text{goal}}^i}^i$; $\quad\quad$ *// assign new goal location*
25: **end if**
$\quad$ *Determine the new requested area for all robot $i \in W_{\mathcal{M}_{\text{on}}}^{\blacklozenge}$*
26: **for each** $i \in W_{\mathcal{M}_{\text{on}}}^{\blacklozenge}$ **do**
27: $\quad\quad$ $\mathbf{B}_{i,\text{estimate}} \leftarrow \text{convexHull}(\mathbf{x}^i \rightarrow W_{\mathbf{q}_{\text{goal}}^i}^{\blacklozenge})$;

28:     $\mathbf{B}_{i,\text{candidate}}$ getAnEstimateOfConvexHull$(\mathbf{B}_{i,\text{estimate}}, W_{\mathbf{A}_i}^{\diamond})$;

29:     $W_{\mathbf{B}_i,\diamond}^{\diamond}$ $\mathbf{B}_{i,\text{candidate}} \cap W_{\text{area}}^{\diamond}$;

30:     $W_{\mathbf{A}_i,\diamond}^{\diamond} \leftarrow W_{\mathbf{A}_i,\diamond}^{\diamond} \cap W_{\mathbf{B}_i,\diamond}^{\diamond}$;

31:     **if**$(\mathbf{B}_{i,\text{candidate}} \not\subseteq W_{\mathbf{B}_i}^{\diamond})$

32:        $W_{\text{counter},i}^{\diamond} \leftarrow W_{\text{counter},i}^{\diamond} + 1$;

33:     **end if**

34:     $W_{\mathbf{B}_i}^{\diamond} \leftarrow \mathbf{B}_{i,\text{candidate}}$;

35: **end for**

    Update the local reserved areas on the agent $\diamond$

36: **for each** $i \in W_{\mathcal{M}}^{\diamond}$ **do**

37:     initialize $W_{\mathbf{A}_i,\diamond,\text{avoided}}^{\diamond}$;

38:     **for each** $j \in W_{\mathcal{M}}^{\diamond} \wedge j \neq i$ **do**

39:        **if** $(W_{\text{cost},i}^{\diamond} < W_{\text{cost},j}^{\diamond})$

40:           $W_{\mathbf{A}_i,\diamond,\text{avoided}}^{\diamond} \leftarrow W_{\mathbf{A}_i,\diamond,\text{avoided}}^{\diamond} \cup W_{\mathbf{A}_j,\diamond}^{\diamond}$;

41:        **else**

42:           $W_{\mathbf{A}_i,\diamond,\text{avoided}}^{\diamond} \leftarrow W_{\mathbf{A}_i,\diamond,\text{avoided}}^{\diamond} \cup W_{\mathbf{B}_j,\diamond}^{\diamond}$;

43:        **end if**

44:     **end for**

45:     $W_{\mathbf{A}_i,\diamond}^{\diamond} = W_{\mathbf{A}_i,\diamond}^{\diamond} \cup (W_{\mathbf{B}_i,\diamond}^{\diamond} \setminus W_{\mathbf{A}_i,\diamond,\text{avoided}}^{\diamond})$;

46: **end for**

    Update the total reserved areas on the whole working environment

47: **for each** $i \in W_{\mathcal{M}_{\text{on}}}^{\diamond}$ **do**

48:     **if**$(\text{isValid}(P^{\clubsuit}))$ *// check if the message is valid according to the information update scheme*

49:        $\mathbf{A}_i^{\clubsuit}$ $W_{\mathbf{B}_i,\diamond}^{\diamond} \cap P_{\mathbf{A}_i,\clubsuit}^{\diamond}$;

50:        $W_{\mathbf{A}_i}^{\diamond}$ $(W_{\mathbf{A}_i}^{\diamond} \setminus (W_{\text{area}}^{\diamond} \cup W_{\text{area}}^{\clubsuit})) \cup W_{\mathbf{A}_i,\diamond}^{\diamond} \cup \mathbf{A}_i^{\clubsuit}$;

51:     **else**

52:        $W_{\mathbf{A}_i}^{\diamond}$ $(W_{\mathbf{A}_i}^{\diamond} \setminus W_{\text{area}}^{\diamond}) \cup W_{\mathbf{A}_i,\diamond}^{\diamond}$;

53:     **end if**

54: **end for**

Subsequently, the reserved areas are proceeded. For each robot $i \in W_{\mathcal{M}}^{\diamond}$, a local avoided area $W_{\mathbf{A}_i,\diamond,\text{avoided}}^{\diamond}$ is computed, which is composed of the local reserved areas $W_{\mathbf{A}_j,\diamond}^{\diamond}$ of lower priority robots and the local requested areas $W_{\mathbf{B}_j,\diamond}^{\diamond}$ of higher priority robots with $i,\ j \in W_{\mathcal{M}}^{\diamond}$ and $i \neq j$ according to lines 35-41. Afterwards, the local reserved area $W_{\mathbf{A}_i,\diamond}^{\diamond}$ is assigned as an union of the previous one and the desired local requested $W_{\mathbf{B}_i,\diamond}^{\diamond}$ area except the avoided area $W_{\mathbf{A}_i,\diamond,\text{avoided}}^{\diamond}$. For each robot $i \in W_{\mathcal{M}_{\text{on}}}^{\diamond}$ on the agent, the reserved area $W_{\mathbf{A}_i}^{\diamond}$ on the whole working environment is eventually updated depending on the message reserved area $P_{\mathbf{A}_i,\clubsuit}^{\clubsuit}$ as stated in lines 49-50. Finally, collision-free trajectories for the robot i can be generated based on the reserved area $W_{\mathbf{A}_i}^{\diamond}$, which is introduced in the following subsection.

6.5.2 Trajectory Generation Function

As shown in Algorithm 6.2, the trajectory generation function consists of two consecutive steps. Primarily, trajectory segments $W_{S^i}^{\blacklozenge}$ are generated for each robot i placed on the agent. Particularly, the robot i occupied area by following the trajectory segments must be inside the reserved area $W_{A_i}^{\blacklozenge}$. In the implementation, a temporary goal is iteratively searched, which is placed on a line connecting the robot's current and goal locations. Moreover, the temporary goal should be as close as possible to the goal location. As aforementioned, a generated trajectory always lies within the convex hulls formed by the control vertices of the quintic Bézier splines. By utilizing this property, one can verify efficiently if the robot area occupied by the trajectory to the temporary goal resides in the reserved area $W_{A_i}^{\blacklozenge}$.

Algorithm 6.2 Trajectory generation function for an agent $\blacklozenge$

    Input: local agent state variables $W^{\blacklozenge}$ and robot state $\mathbf{x}^i$

    Output: robot trajectory segment $W_{S^i}^{\blacklozenge}$ and new agent state variables $W^{\blacklozenge}$

1: **for each** $i \in W_{\mathcal{M}_{on}}^{\blacklozenge}$ **do**

2:      $W_{S^i}^{\blacklozenge}$ generateTrajectorySegment$(W_{A_i}^{\blacklozenge}, \mathbf{x}^i(t), W_{\mathbf{q}_{goal}^i}^{\blacklozenge})$;

3:      $W^{\blacklozenge}$ updateNewState$(W^{\blacklozenge})$;

4: **end for**

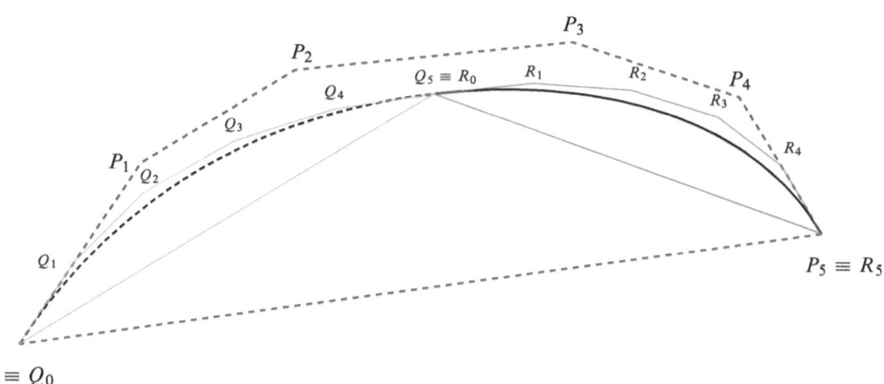

Figure 6.5: Subdividing a quintic Bézier spline.

Subsequently, the agent state $W^{\blacklozenge}$ is updated. As mentioned in Section 6.3, the reserved and requested areas will be decreased as long as the robots follow their trajectories. However, these areas can be efficiently computed by exploiting a Bézier spline's property. As shown in Figure 6.5, the shrinking areas can be estimated without any complex geometrical operation by using the subdivision De Casteljau algorithm [26].

Particularly, a quintic Bézier spline is defined by

$$
\mathbf{q}(u) = \underbrace{(1-u)^5\, P_0}_{B_0(u)} + \underbrace{5(1-u)^4 u\, P_1}_{B_1(u)} + \underbrace{10(1-u)^3 u^2\, P_2}_{B_2(u)}
$$

$$
+ \underbrace{10(1-u)^2 u^3\, P_3}_{B_3(u)} + \underbrace{5(1-u)u^4\, P_4}_{B_4(u)} + \underbrace{u^5}_{B_5(u)}\, P_5
$$

$$
= \sum_{i=0}^{n=5} P_i B_i(u), \quad u \in [0 \quad 1] \quad \text{and} \quad n = 5,
$$

where B_i are the Bernstein polynomials and $\mathbf{P} := \{P_0, P_1, \ldots P_5\}$ are the control points. The quintic Bézier spline is split into two Bézier splines at any point $\mathbf{q}(u^*)$ with $u^* \in [0\ 1]$. The control points of the two new Bézier splines can be determined by using the De Casteljau's algorithm. In fact, a sequence of points is given by using a recurrence relation as follows:

$$
P_i^0 := P_i, \qquad i = 0, \ldots, n = 5,
$$

$$
P_i^j := P_i^{(j-1)}(1 - u^*) + P_{i+1}^{(j-1)} u^*, \qquad i = 0, \ldots, n - j, \quad j = 1, \ldots, n.
$$

The two new Bézier splines are determined by the following control points

$$
\mathbf{Q} := \{Q_0,\, Q_1,\, \ldots,\, Q_n\} := \{P_0^0,\, P_0^1,\, \ldots,\, P_0^n\},
$$

$$
\mathbf{R} := \{R_0,\, R_1,\, \ldots,\, R_n\} := \{P_0^n,\, P_1^{n-1},\, \ldots,\, P_n^0\}.
$$

As shown in Figure 6.5, the dashed line represents the trajectory segment, which has been executed and is inside the convex hull $\mathrm{Conv}(\mathbf{Q})$. On the other hand, the rest of the spline for $u \in [u^*\ \ 1]$ lies within the convex hull $\mathrm{Conv}(\mathbf{R})$. Furthermore, the convex hull $\mathrm{Conv}(\mathbf{R})$ is a subset of the original convex hull $\mathrm{Conv}(\mathbf{P})$. Therefore, the convex hull $\mathrm{Conv}(\mathbf{R})$ can be used as an estimate of the shrinking reserved and requested areas.

6.5.3 Message Generation Function

Algorithm 6.3 Message generation function for an agent $\blacklozenge$

    Input: local agent state variables $W^{\blacklozenge}$ and robot state $\mathbf{x}^i$

    Output: a list of out messages $\mathbf{P}_{\text{out}}$

1: **for each** $i \in W_{\mathcal{M}}^{\blacklozenge}$ **do**

2:      $\tilde{P}_{id} \quad \blacklozenge$;

3:      $\tilde{P}_{t_{\text{last}}} \quad W_{t_{\text{local,clock}}}^{\blacklozenge}$;

4:      $\tilde{P}_i \quad i$;

5:      $\tilde{P}_{\text{counter},i} \quad W_{\text{counter},i}^{\blacklozenge}$;

6:      $\tilde{P}_{\mathbf{A}_i,\blacklozenge} \quad W_{\mathbf{A}_i,\blacklozenge}^{\blacklozenge}$;

7:      **if** $(i \in W_{\mathcal{M}_{\text{on}}}^{\blacklozenge})$

8:          $\tilde{P}_{\mathbf{B}_i} \quad W_{\mathbf{B}_i}^{\blacklozenge}$;

9:          $\tilde{P}_{\mathbf{x}^i} \quad \mathbf{x}^i$;

10:      **end if**
11:          $\mathbf{P}_{\text{out}} \leftarrow \mathbf{P}_{\text{out}} \cup \{\tilde{P}\}$;
12: **end for**

As stated in Algorithm 6.3, a message is generated and sent to other related agents for each robot $i \in W_{\mathcal{M}}^{\diamondsuit}$. Particularly, all necessary information given in Table 6.2 are compressed in the message.

6.6 A Formal Safety Guarantee Proof

In spite of stochastic communication time delays among agents, the invariant safety can be guaranteed by the distributed collision avoidance algorithm. Particularly, if the reserved areas are distinct at the algorithm initialization, then the invariant safety guarantee can be formally stated as the following theorem.

Theorem 6.1 *Assume that communication time delays among agents are greater than zero ($\Delta T_d > 0$). If the following conditions*

$$W_{\mathbf{A}_i}^{\diamondsuit}(t) \cap W_{\mathbf{A}_j}^{\spadesuit}(t) = \varnothing, \quad \forall i \neq j, \quad i \in W_{\mathcal{M}_{on}}^{\diamondsuit} \wedge j \in W_{\mathcal{M}_{on}}^{\spadesuit} \tag{6.1}$$

$$W_{\mathbf{A}_i}^{\diamondsuit}(t) \cap W_{\mathbf{A}_j, \spadesuit}^{\spadesuit}(t) = \varnothing, \quad \forall i \neq j, \quad \forall\ \diamondsuit, \spadesuit \tag{6.2}$$

$$W_{\mathbf{A}_i, \diamondsuit}^{\diamondsuit}(t) \cap W_{\mathbf{A}_j, \diamondsuit}^{\diamondsuit}(t) = \varnothing, \quad \forall i \neq j, \quad \forall\ \diamondsuit, \tag{6.3}$$

hold for $t \leq t_0$, where t_0 is the initialization time, then the distributed collision avoidance algorithm is invariantly safe.

Proof As aforementioned, each agent $\diamondsuit$ only plans trajectories for a robot i, which is placed on the agent $\diamondsuit$, i.e. $i \in W_{\mathcal{M}_{on}}^{\diamondsuit}$. Additionally, the trajectories are generated such that the robot occupied area is always inside the local reserved area $W_{\mathbf{A}_i}^{\diamondsuit}$. Therefore, if the condition

$$W_{\mathbf{A}_i}^{\diamondsuit}(t) \cap W_{\mathbf{A}_j}^{\spadesuit}(t) = \varnothing, \quad \forall i \neq j, \quad i \in W_{\mathcal{M}_{on}}^{\diamondsuit} \wedge j \in W_{\mathcal{M}_{on}}^{\spadesuit}, \tag{6.4}$$

holds at any time instant t, then the distributed collision avoidance algorithm is invariantly safe.

By using the induction principle, one is able show that the condition (6.4) is satisfied at any time instant t. From the algorithm initialization, one has that

$$W_{\mathbf{A}_i}^{\diamondsuit}(t) \cap W_{\mathbf{A}_j}^{\spadesuit}(t) = \varnothing, \quad \forall i \neq j, \quad i \in W_{\mathcal{M}_{on}}^{\diamondsuit} \wedge j \in W_{\mathcal{M}_{on}}^{\spadesuit}, \tag{6.5}$$

holds for any time instant $t \leq t_0$. Let t_e is defined as a time stamp when one of the following areas $W_{\mathbf{B}_i}^{\diamondsuit}$, $W_{\mathbf{B}_j}^{\spadesuit}$, $W_{\mathbf{A}_i, \diamondsuit}^{\diamondsuit}$, $W_{\mathbf{A}_j, \spadesuit}^{\spadesuit}$ are changed. Moreover, the time-stamp t_e can be defined, when either the agent $\blacklozenge$ receives a valid update message $P^{\clubsuit}$ for the robot i or the agent $\spadesuit$ receives a valid update message $P^{\star}$ for the robot j. If one is able to show that

$$W_{\mathbf{A}_i}^{\blacklozenge}(t) \cap W_{\mathbf{A}_j}^{\spadesuit}(t) = \varnothing, \quad \forall t_0 < t \leq t_e, \quad \forall i \neq j, \quad i \in W_{\mathcal{M}_{on}}^{\blacklozenge} \wedge j \in W_{\mathcal{M}_{on}}^{\spadesuit}, \qquad (6.6)$$

then t_e can be reset as t_0 in the next time interval. Consequently, the condition (6.4) holds for any time instant t and the distributed collision avoidance algorithm is invariantly safe by the mean of the induction principle.

According to lines 49-50 and 52 of Algorithm 6.1, the robot i reserved area $W_{\mathbf{A}_i}^{\blacklozenge}$ is updated by either

$$W_{\mathbf{A}_i}^{\blacklozenge}(t) := \left(W_{\mathbf{A}_i}^{\blacklozenge}(t_0) \setminus (W_{\text{area}}^{\blacklozenge} \cup W_{\text{area}}^{\clubsuit})\right) \cup W_{\mathbf{A}_{i,\blacklozenge}}^{\blacklozenge}(t_0) \cup \mathbf{A}_i^{\clubsuit}(t_0) \quad \text{or} \qquad (6.7)$$

$$W_{\mathbf{A}_i}^{\blacklozenge}(t) := \left(W_{\mathbf{A}_i}^{\blacklozenge}(t_0) \setminus W_{\text{area}}^{\blacklozenge}\right) \cup W_{\mathbf{A}_{i,\blacklozenge}}^{\blacklozenge}(t_0), \quad \forall t_0 < t \leq t_e. \qquad (6.8)$$

Analogously, the robot j reserved area $W_{\mathbf{A}_j}^{\spadesuit}$ is updated by either

$$W_{\mathbf{A}_j}^{\spadesuit}(t) := \left(W_{\mathbf{A}_j}^{\spadesuit}(t_0) \setminus (W_{\text{area}}^{\spadesuit} \cup W_{\text{area}}^{\star})\right) \cup W_{\mathbf{A}_{j,\spadesuit}}^{\spadesuit}(t_0) \cup \mathbf{A}_j^{\star}(t_0) \quad \text{or} \qquad (6.9)$$

$$W_{\mathbf{A}_j}^{\spadesuit}(t) := \left(W_{\mathbf{A}_j}^{\spadesuit}(t_0) \setminus W_{\text{area}}^{\spadesuit}\right) \cup W_{\mathbf{A}_{j,\spadesuit}}^{\spadesuit}(t_0), \quad \forall t_0 < t \leq t_e. \qquad (6.10)$$

From (6.7), (6.8), (6.9) and (6.10), there are four cases in computing the intersection of the robot i and j reserved areas. Nevertheless, the most general case is a combination of (6.7) and (6.9). By substituting (6.7) and (6.9) into (6.6), one consequently has to show that

$$
\begin{aligned}
W_{\mathbf{A}_i}^{\blacklozenge}(t) \cap W_{\mathbf{A}_j}^{\spadesuit}(t) = {} & \left(\left(W_{\mathbf{A}_i}^{\blacklozenge}(t_0) \setminus (W_{\text{area}}^{\blacklozenge} \cup W_{\text{area}}^{\clubsuit})\right) \cup W_{\mathbf{A}_{i,\blacklozenge}}^{\blacklozenge}(t_0) \cup \mathbf{A}_i^{\clubsuit}(t_0)\right) \\
& \cap \left(\left(W_{\mathbf{A}_j}^{\spadesuit}(t_0) \setminus (W_{\text{area}}^{\spadesuit} \cup W_{\text{area}}^{\star})\right) \cup W_{\mathbf{A}_{j,\spadesuit}}^{\spadesuit}(t_0) \cup \mathbf{A}_j^{\star}(t_0)\right) \\
= {} & \underbrace{\left[\left(W_{\mathbf{A}_i}^{\blacklozenge}(t_0) \setminus (W_{\text{area}}^{\blacklozenge} \cup W_{\text{area}}^{\clubsuit})\right) \cap \left(W_{\mathbf{A}_j}^{\spadesuit}(t_0) \setminus (W_{\text{area}}^{\spadesuit} \cup W_{\text{area}}^{\star})\right)\right]}_{T_1} \cup \\
& \underbrace{\left[W_{\mathbf{A}_{j,\spadesuit}}^{\spadesuit}(t_0) \cap \left(W_{\mathbf{A}_i}^{\blacklozenge}(t_0) \setminus (W_{\text{area}}^{\blacklozenge} \cup W_{\text{area}}^{\clubsuit})\right)\right]}_{T_2} \cup \\
& \underbrace{\left[\left(W_{\mathbf{A}_i}^{\blacklozenge}(t_0) \setminus (W_{\text{area}}^{\blacklozenge} \cup W_{\text{area}}^{\clubsuit})\right) \cap \mathbf{A}_j^{\star}(t_0)\right]}_{T_3} \cup \\
& \underbrace{\left[W_{\mathbf{A}_{i,\blacklozenge}}^{\blacklozenge}(t_0) \cap \left(W_{\mathbf{A}_j}^{\spadesuit}(t_0) \setminus (W_{\text{area}}^{\spadesuit} \cup W_{\text{area}}^{\star})\right)\right]}_{T_4} \cup \\
& \underbrace{\left[W_{\mathbf{A}_{i,\blacklozenge}}^{\blacklozenge}(t_0) \cap W_{\mathbf{A}_{j,\spadesuit}}^{\spadesuit}(t_0)\right]}_{T_5} \cup \underbrace{\left[W_{\mathbf{A}_{i,\blacklozenge}}^{\blacklozenge}(t_0) \cap \mathbf{A}_j^{\star}(t_0)\right]}_{T_6} \cup \\
& \underbrace{\left[\mathbf{A}_i^{\clubsuit}(t_0) \cap \left(W_{\mathbf{A}_j}^{\spadesuit}(t_0) \setminus (W_{\text{area}}^{\spadesuit} \cup W_{\text{area}}^{\star})\right)\right]}_{T_7} \cup
\end{aligned}
\qquad (6.11)
$$

$$\underbrace{\left[\mathbf{A}_i^{\spadesuit}(t_0) \cap W_{\mathbf{A}_j,\blacklozenge}^{\spadesuit}(t_0)\right]}_{T_8} \cup \underbrace{\left[\mathbf{A}_i^{\spadesuit}(t_0) \cap \mathbf{A}_j^{\star}(t_0)\right]}_{T_9}$$

$$= \quad \varnothing, \qquad \forall t_0 < t \le t_e,$$

holds in order to accomplish the proof. In the following, we are able to show that the terms $T_i = \varnothing$, $\forall i \in \{1, \ldots, 9\}$.

- $T_1 = \varnothing$:

For the term T_1, one has

$$
\begin{aligned}
T_1 &= \left(W_{\mathbf{A}_i}^{\blacklozenge}(t_0) \setminus (W_{\text{area}}^{\blacklozenge} \cup W_{\text{area}}^{\clubsuit})\right) \cap \left(W_{\mathbf{A}_j}^{\spadesuit}(t_0) \setminus (W_{\text{area}}^{\spadesuit} \cup W_{\text{area}}^{\star})\right) \\
&= \underbrace{\left(W_{\mathbf{A}_i}^{\blacklozenge}(t_0) \cap W_{\mathbf{A}_j}^{\spadesuit}(t_0)\right)}_{=\varnothing} \setminus (W_{\text{area}}^{\blacklozenge} \cup W_{\text{area}}^{\clubsuit} \cup W_{\text{area}}^{\spadesuit} \cup W_{\text{area}}^{\star}) \\
&= \varnothing,
\end{aligned}
\tag{6.12}
$$

since one has from the initialization (6.5) that $W_{\mathbf{A}_i}^{\blacklozenge}(t_0) \cap W_{\mathbf{A}_j}^{\spadesuit}(t_0) = \varnothing$, $\forall t \le t_0$.

- $T_2 = T_4 = \varnothing$:

By interchanging the agents $\blacklozenge$ and $\spadesuit$, the robots i and j, the sending message agents $\star$ and $\clubsuit$, one has that the following term

$$
\begin{aligned}
W_{\mathbf{A}_j,\spadesuit}^{\spadesuit}(t_0) &\equiv W_{\mathbf{A}_i,\blacklozenge}^{\blacklozenge}(t_0) \\
W_{\mathbf{A}_i}^{\blacklozenge}(t_0) &\equiv W_{\mathbf{A}_j}^{\spadesuit}(t_0) \\
W_{\text{area}}^{\blacklozenge} &\equiv W_{\text{area}}^{\spadesuit} \\
W_{\text{area}}^{\clubsuit} &\equiv W_{\text{area}}^{\star},
\end{aligned}
$$

are equivalent. From (6.11), the term

$$T_2 = W_{\mathbf{A}_j,\spadesuit}^{\spadesuit}(t_0) \cap \left(W_{\mathbf{A}_i}^{\blacklozenge}(t_0) \setminus (W_{\text{area}}^{\blacklozenge} \cup W_{\text{area}}^{\clubsuit})\right),$$

which is obviously symmetrical to the term

$$T_4 = W_{\mathbf{A}_i,\blacklozenge}^{\blacklozenge}(t_0) \cap \left(W_{\mathbf{A}_j}^{\spadesuit}(t_0) \setminus (W_{\text{area}}^{\spadesuit} \cup W_{\text{area}}^{\star})\right)$$

since all four elements are equivalent. Therefore, it is sufficient to show that the term $T_2 = \varnothing$. By reformulating the T_2, one obtains that

$$
\begin{aligned}
T_2 &= W_{\mathbf{A}_j,\spadesuit}^{\spadesuit}(t_0) \cap \left(W_{\mathbf{A}_i}^{\blacklozenge}(t_0) \setminus (W_{\text{area}}^{\blacklozenge} \cup W_{\text{area}}^{\clubsuit})\right) \\
&= \left(W_{\mathbf{A}_i}^{\blacklozenge}(t_0) \cap W_{\mathbf{A}_j,\spadesuit}^{\spadesuit}(t_0)\right) \setminus (W_{\text{area}}^{\blacklozenge} \cup W_{\text{area}}^{\clubsuit}).
\end{aligned}
\tag{6.13}
$$

In the following, two following cases are considered. Particularly, if the agents $\blacklozenge$ and $\spadesuit$ are either identical or distinct.

Case 1: If $\blacklozenge \equiv \spadesuit$, then the term $T_2 = \varnothing$ is obviously true, since the right side of the set minus $W_{\text{area}}^{\blacklozenge=\spadesuit} \cup W_{\text{area}}^{\clubsuit}$ contains the local area $W_{\mathbf{A}_j,\spadesuit}^{\spadesuit}(t_0)$, i.e.

$$W_{\mathbf{A}_i}^{\blacklozenge}(t_0) \cap W_{\mathbf{A}_j,\spadesuit}^{\spadesuit}(t_0) \subseteq W_{\mathbf{A}_j,\spadesuit}^{\spadesuit}(t_0) \subseteq W_{\text{area}}^{\blacklozenge=\spadesuit} \subseteq W_{\text{area}}^{\blacklozenge=\spadesuit} \cup W_{\text{area}}^{\clubsuit}.$$

Case 2: If $\blacklozenge \neq \spadesuit$, one is also able to show

$$W_{\mathbf{A}_i}^{\blacklozenge}(t) \cap W_{\mathbf{A}_j,\spadesuit}^{\spadesuit}(t) = \varnothing, \quad \forall i \neq j, \quad \forall \blacklozenge, \spadesuit, \quad \forall\, t, \tag{6.14}$$

by using the induction method again. Moreover, one has from the initialization (6.2) that

$$W_{\mathbf{A}_i}^{\blacklozenge}(t) \cap W_{\mathbf{A}_j,\spadesuit}^{\spadesuit}(t) = \varnothing, \quad \forall i \neq j, \quad \forall \blacklozenge, \spadesuit, \quad \forall\, t \leq t_0. \tag{6.15}$$

Let denote t_1 as a time stamp, when either $W_{\mathbf{A}_i}^{\blacklozenge}(t)$ or $W_{\mathbf{A}_j,\spadesuit}^{\spadesuit}(t)$ is changed. One has to show that

$$W_{\mathbf{A}_i}^{\blacklozenge}(t) \cap W_{\mathbf{A}_j,\spadesuit}^{\spadesuit}(t) = \varnothing, \quad \forall\, t_0 < t \leq t_1. \tag{6.16}$$

Moreover, t_1 can be reset as t_0 of the next time interval. By the mean of the induction principle, one has that the condition (6.14) holds. As stated in lines 49-50 and 52 of Algorithm 6.1, the total reserved area $W_{\mathbf{A}_i}^{\blacklozenge}(t)$ can be updated by either

$$W_{\mathbf{A}_i}^{\blacklozenge}(t) = \left(W_{\mathbf{A}_i}^{\blacklozenge}(t_0) \setminus (W_{\text{area}}^{\blacklozenge} \cup W_{\text{area}}^{\clubsuit})\right) \cup W_{\mathbf{A}_{i,\blacklozenge}}^{\blacklozenge}(t_0) \cup \mathbf{A}_i^{\clubsuit}(t_0) \quad \text{or} \tag{6.17}$$

$$W_{\mathbf{A}_i}^{\blacklozenge}(t) = (W_{\mathbf{A}_i}^{\blacklozenge}(t_0) \setminus W_{\text{area}}^{\blacklozenge}) \cup W_{\mathbf{A}_{i,\blacklozenge}}^{\blacklozenge}(t_0), \quad \forall t_0 < t \leq t_1. \tag{6.18}$$

Furthermore, the update (6.17) is more general than (6.18). Therefore, it is sufficient to consider the update (6.17). By substituting (6.17) into (6.16), one has that the intersection

$$
\begin{aligned}
W_{\mathbf{A}_i}^{\blacklozenge}(t) &\cap W_{\mathbf{A}_j,\spadesuit}^{\spadesuit}(t) \\
&= \left[\left(W_{\mathbf{A}_i}^{\blacklozenge}(t_0) \setminus (W_{\text{area}}^{\blacklozenge} \cup W_{\text{area}}^{\clubsuit})\right) \cup W_{\mathbf{A}_{i,\blacklozenge}}^{\blacklozenge}(t_0) \cup \mathbf{A}_i^{\clubsuit}(t_0)\right] \cap W_{\mathbf{A}_j,\spadesuit}^{\spadesuit}(t_0) \\
&= \underbrace{\left[\left(W_{\mathbf{A}_i}^{\blacklozenge}(t_0) \cap W_{\mathbf{A}_j,\spadesuit}^{\spadesuit}(t_0)\right) \setminus (W_{\text{area}}^{\blacklozenge} \cup W_{\text{area}}^{\clubsuit})\right]}_{T_1^*} \cup \\
&\quad \underbrace{\left[W_{\mathbf{A}_{i,\blacklozenge}}^{\blacklozenge}(t_0) \cap W_{\mathbf{A}_j,\spadesuit}^{\spadesuit}(t_0)\right]}_{T_2^*} \cup \underbrace{\left[\mathbf{A}_i^{\clubsuit}(t_0) \cap W_{\mathbf{A}_j,\spadesuit}^{\spadesuit}(t_0)\right]}_{T_3^*}.
\end{aligned}
\tag{6.19}
$$

From the initialization (6.15), the left side of the term T_1^* is an empty set. Consequently, one has that

$$T_1^* = \left(W_{\mathbf{A}_i}^{\blacklozenge}(t_0) \cap W_{\mathbf{A}_j,\spadesuit}^{\spadesuit}(t_0)\right) \setminus (W_{\text{area}}^{\blacklozenge} \cup W_{\text{area}}^{\clubsuit}) = \varnothing.$$

Moreover, the intersection of the local reserved areas from different agents

$$T_2^* = W_{\mathbf{A}_{i,\blacklozenge}}^{\blacklozenge}(t_0) \cap W_{\mathbf{A}_j,\spadesuit}^{\spadesuit}(t_0) \subseteq W_{\text{area}}^{\blacklozenge} \cap W_{\text{area}}^{\spadesuit} = \varnothing, \quad \spadesuit \neq \blacklozenge, \Rightarrow T_2^* = \varnothing,$$

is obviously an empty set. For the term T_3^*, the following subcases are considered.

Subcase ♣ ≢ ♠:

The reserved area $\mathbf{A}_i^{♣}$ is only a subset of the agent area $W_{\text{area}}^{♣}$. Similarly, the local reserved area $W_{\mathbf{A}_j,♠}^{♠}$ is also a subset of the agent area $W_{\text{area}}^{♠}$. Since ♣ ≢ ♠, one has $W_{\text{area}}^{♣} \cap W_{\text{area}}^{♠} = \varnothing$, which consequently leads to

$$T_3^* = \mathbf{A}_i^{♣}(t_0) \cap W_{\mathbf{A}_j,♠}^{♠}(t_0) \subseteq W_{\text{area}}^{♣} \cap W_{\text{area}}^{♠} = \varnothing. \tag{6.20}$$

Subcase ♣ ≡ ♠:

If ♣ ≡ ♠, $T_3^* = \varnothing$ can be shown in the following. Since the message $P^{♣}$ is valid, there is at least one successful communication cycle between the agent ♣ and the agents ♦. Moreover, the milestone time-stamps of the last communication cycle are depicted in Figure 6.6. Let assume that the agent ♦ receives the message $P^{♣}$ at the time-stamp t_0. From line 49 of Algorithm 6.1, one has

$$\mathbf{A}_i^{♦}(t_0) = W_{\mathbf{B}_i}^{♦}(t_0) \cap P_{\mathbf{A}_i,♣}^{♣}(t_0). \tag{6.21}$$

As shown in Figure 6.6, we additionally have that

$$P_{\mathbf{A}_i,♣}^{♣}(t_0) = W_{\mathbf{A}_i,♣}^{♣}(t_0^*), \tag{6.22}$$

where t_0^* is the time stamp, when the agent ♣ sent the message $P^{♣}$ to the agent ♦. Moreover, t_{-2}^* denotes the time-stamp, when the agent ♦ initialized a new communication cycle due to enlargement of the requested area $W_{\mathbf{B}_i}^{♦}$ by sending a message $P^{♦}$. As stated in lines 31-32 of Algorithm 6.1, the state variable $W_{\text{counter},i}^{♦}$ is consequently increased. Furthermore, the message $P^{♦}$ is received by the agent ♣ at the time-stamp t_{-1}^*. During the time interval between t_{-2}^* and t_0, a message $\tilde{P}^{♦}$ can be sent by the agent ♦, indicating $i \in W_{\mathcal{M}_{\text{on}}}^{♦}(\tilde{t}_{-2})$ at the time-stamp $\tilde{t}_{-2}$. As illustrated in Figure 6.6, the message $\tilde{P}^{♦}$ is received by the agent ♣ at a time-stamp $\tilde{t}_{-1}$. Since communication delays are greater than zero, one consequently has $t_{-2}^* < t_{-1}^* \leq t_0^* < t_0, t_{-2}^* \leq \tilde{t}_{-2}$, $t_0^* \leq \tilde{t}_{-1}$ and $\tilde{t}_{-2} < \tilde{t}_{-1}$.

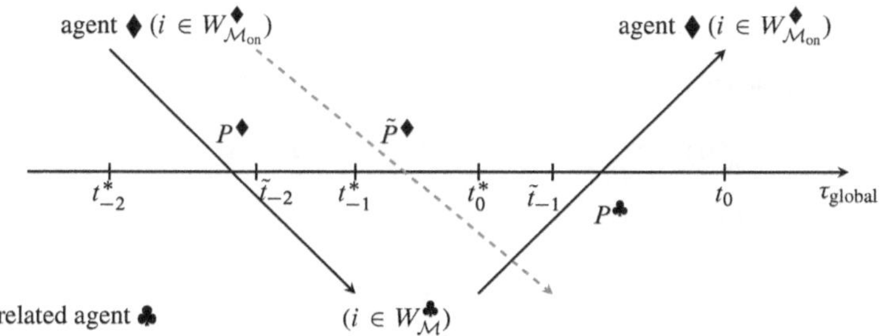

Figure 6.6: An information update scheme for a robot i

Substituting (6.22) into (6.21) leads to

$$\mathbf{A}_i^{♦}(t_0) = W_{\mathbf{B}_i}^{♦}(t_0) \cap W_{\mathbf{A}_i,♣}^{♣}(t_0^*). \tag{6.23}$$

From lines 6-8 and lines 40-42 of Algorithm 6.1, we have that

$$
\begin{aligned}
W_{A_i,\clubsuit}^{\clubsuit}(t_0) &= \left[W_{A_i,\clubsuit}^{\clubsuit}(t_0^*) \cap \left(W_{B_i}^{\diamondsuit}(\tilde{t}_{-2}) \cap W_{\text{area}}^{\clubsuit} \right) \right] \cup \left[W_{B_i,\clubsuit}^{\clubsuit}(\tilde{t}_{-1}) \setminus W_{A_i,\clubsuit,\text{avoided}}^{\clubsuit}(\tilde{t}_{-1}) \right] \\
&\supseteq \left[W_{A_i,\clubsuit}^{\clubsuit}(t_0^*) \cap W_{B_i}^{\diamondsuit}(\tilde{t}_{-2}) \cap W_{\text{area}}^{\clubsuit} \right] \\
&= \underbrace{W_{A_i,\clubsuit}^{\clubsuit}(t_0^*) \cap W_{\text{area}}^{\clubsuit}}_{=W_{A_i,\clubsuit}^{\clubsuit}(t_0^*),\ \text{since}\ W_{A_i,\clubsuit}^{\clubsuit}(t_0^*) \subseteq W_{\text{area}}^{\clubsuit}} \cap W_{B_i}^{\diamondsuit}(\tilde{t}_{-2}) \\
&= W_{A_i,\clubsuit}^{\clubsuit}(t_0^*) \cap W_{B_i}^{\diamondsuit}(\tilde{t}_{-2}).
\end{aligned}
\tag{6.24}
$$

Since the message $P^{\clubsuit}$ is accepted, one has that

$$
W_{B_i}^{\diamondsuit}(\tilde{t}_{-2}) \supseteq W_{B_i}^{\diamondsuit}(t_0).
\tag{6.25}
$$

Otherwise, it will result in $W_{\text{counter},i}^{\diamondsuit} > P_{\text{counter},i}^{\diamondsuit}$, which immediately leads to a contradiction that the message $P^{\clubsuit}$ is accepted. By substituting (6.22) into (6.21) and from (6.25), (6.24), one obtains

$$
\begin{aligned}
A_i^{\clubsuit}(t_0) &= W_{B_i}^{\diamondsuit}(t_0) \cap W_{A_i,\clubsuit}^{\clubsuit}(t_0^*) \subseteq W_{B_i}^{\diamondsuit}(\tilde{t}_{-2}) \cap W_{A_i,\clubsuit}^{\clubsuit}(t_0^*) \\
&\subseteq W_{A_i,\clubsuit}^{\clubsuit}(t_0)
\end{aligned}
\tag{6.26}
$$

Eventually, we have from (6.26) and (6.19) that

$$
T_3^* = A_i^{\clubsuit}(t_0) \cap W_{A_j,\clubsuit}^{\clubsuit}(t_0) \subseteq W_{A_i,\clubsuit}^{\clubsuit}(t_0) \cap W_{A_j,\clubsuit}^{\clubsuit}(t_0).
\tag{6.27}
$$

In order to prove that $T_5 = \varnothing$, we will show later in (6.46) that

$$
W_{A_i,\clubsuit}^{\clubsuit}(t) \cap W_{A_j,\clubsuit}^{\clubsuit}(t) = \varnothing, \quad \forall\, i \neq j, \quad \forall\, t, \quad \forall\, \clubsuit.
\tag{6.28}
$$

From (6.27) and (6.28), it leads to

$$
T_3^* = \varnothing.
\tag{6.29}
$$

Finally, we have shown that $T_1^* = T_2^* = T_3^* = \varnothing$. Consequently, one has that

$$
W_{A_i}^{\diamondsuit}(t) \cap W_{A_j,\clubsuit}^{\spadesuit}(t) = \varnothing, \quad \forall i \neq j, \quad \forall\, \diamondsuit, \spadesuit, \quad \forall\, t.
\tag{6.30}
$$

By replacing (6.30) into (6.13), one has

$$
T_2 = T_4 = \varnothing.
\tag{6.31}
$$

- $T_3 = T_7 = \varnothing$

In analogy to the terms T_2 and T_4, the terms T_3 and T_7 are also symmetrical by interchanging the agent and robot indexes. Consequently, we only have to show that the term $T_3 = \varnothing$. For the term T_3, one has

$$
T_3 = \left(W_{A_i}^{\diamondsuit}(t_0) \setminus \left(W_{\text{area}}^{\diamondsuit} \cup W_{\text{area}}^{\clubsuit} \right) \right) \cap A_j^{\star}(t_0) \subseteq W_{A_i}^{\diamondsuit}(t_0) \cap A_j^{\star}(t_0).
\tag{6.32}
$$

In analogy to (6.26), one has

$$\mathbf{A}_j^{\star}(t_0) \subseteq W_{\mathbf{A}_j,\star}^{\star}(t_0), \quad \forall\, t_0. \tag{6.33}$$

Substituting (6.32) into (6.33) leads to

$$T_3 \subseteq W_{\mathbf{A}_i}^{\blacklozenge}(t_0) \cap W_{\mathbf{A}_j,\star}^{\star}(t_0). \tag{6.34}$$

We have shown in (6.14) that

$$W_{\mathbf{A}_i}^{\blacklozenge}(t) \cap W_{\mathbf{A}_j,\star}^{\star}(t) = \varnothing, \quad \forall\, t, \quad \forall\, i \neq j, \quad \forall\, \star, \blacklozenge, \tag{6.35}$$

which consequently implies to that $T_3 = T_7 = \varnothing$.

- $T_5 = \varnothing$

From (6.11), the term T_5 is defined as

$$T_5 = W_{\mathbf{A}_i,\blacklozenge}^{\blacklozenge}(t_0) \cap W_{\mathbf{A}_j,\spadesuit}^{\spadesuit}(t_0), \quad i \neq j, \quad \forall\, \blacklozenge, \spadesuit. \tag{6.36}$$

In order to show that $T_5 = \varnothing$ holds for every time instant t, the following cases are considered.

Case 1: If $\blacklozenge \not\equiv \spadesuit$, one has that $W_{\text{area}}^{\blacklozenge} \cap W_{\text{area}}^{\spadesuit} = \varnothing$. Moreover, it is obviously that $W_{\mathbf{A}_i,\blacklozenge}^{\blacklozenge}(t) \subseteq W_{\text{area}}^{\blacklozenge}$ and $W_{\mathbf{A}_i,\spadesuit}^{\spadesuit}(t) \subseteq W_{\text{area}}^{\spadesuit}$, which obviously leads to

$$T_5 = W_{\mathbf{A}_i,\blacklozenge}^{\blacklozenge}(t) \cap W_{\mathbf{A}_i,\spadesuit}^{\spadesuit}(t) \subseteq W_{\text{area}}^{\blacklozenge} \cap W_{\text{area}}^{\spadesuit} = \varnothing, \quad \forall t.$$

Case 2: If $\blacklozenge \equiv \spadesuit$, then the induction method can be used again to prove that

$$T_5 = W_{\mathbf{A}_i,\blacklozenge}^{\blacklozenge}(t) \cap W_{\mathbf{A}_j,\spadesuit}^{\spadesuit}(t) = W_{\mathbf{A}_i,\blacklozenge}^{\blacklozenge}(t) \cap W_{\mathbf{A}_j,\blacklozenge}^{\blacklozenge}(t) = \varnothing, \quad \forall\, t, \blacklozenge \equiv \spadesuit, i \neq j. \tag{6.37}$$

From the algorithm initialization (6.3), one has that

$$T_5 = W_{\mathbf{A}_i,\blacklozenge}^{\blacklozenge}(t) \cap W_{\mathbf{A}_j,\blacklozenge}^{\blacklozenge}(t) = \varnothing, \quad \forall\, t \leq t_0, \blacklozenge, i \neq j. \tag{6.38}$$

Moreover, let denote t_2 as a time-stamp, when one of the following areas $W_{\mathbf{B}_i,\blacklozenge}^{\blacklozenge}, W_{\mathbf{B}_j,\blacklozenge}^{\blacklozenge}, W_{\mathbf{A}_i,\blacklozenge,\text{avoided}}^{\blacklozenge}$ and $W_{\mathbf{A}_j,\blacklozenge,\text{avoided}}^{\blacklozenge}$ is changed. Eventually, one has to show that

$$T_5 = W_{\mathbf{A}_i,\blacklozenge}^{\blacklozenge}(t) \cap W_{\mathbf{A}_j,\blacklozenge}^{\blacklozenge}(t) = \varnothing, \quad \forall\, t_0 < t \leq t_2, \blacklozenge, i \neq j. \tag{6.39}$$

In analogy to proving the term $T_2 = \varnothing$, t_2 can be reset as the t_0 of the next time interval. Consequently, the condition $T_5 = \varnothing$ holds by the mean of the induction principle. Furthermore, one has from line 44 of Algorithm 6.1 that

$$\begin{aligned}
W_{\mathbf{A}_i,\blacklozenge}^{\blacklozenge}(t) &= W_{\mathbf{A}_i,\blacklozenge}^{\blacklozenge}(t_0) \cup \left(W_{\mathbf{B}_i,\blacklozenge}^{\blacklozenge}(t_0) \setminus W_{\mathbf{A}_i,\blacklozenge,\text{avoided}}^{\blacklozenge}(t_0) \right), \quad \forall\, t_0 < t \leq t_2, \\
W_{\mathbf{A}_j,\blacklozenge}^{\blacklozenge}(t) &= W_{\mathbf{A}_j,\blacklozenge}^{\blacklozenge}(t_0) \cup \left(W_{\mathbf{B}_j,\blacklozenge}^{\blacklozenge}(t_0) \setminus W_{\mathbf{A}_j,\blacklozenge,\text{avoided}}^{\blacklozenge}(t_0) \right), \quad \forall\, t_0 < t \leq t_2.
\end{aligned} \tag{6.40}$$

Without the loss of generality, let us assume that the robot i has a higher priority than the robot j. Consequently, one also has from lines 39-41 of Algorithm 6.1 that

$$W_{\mathbf{A}_i,\blacklozenge,\text{avoided}}(t_0) \supseteq W_{\mathbf{A}_j,\blacklozenge}(t_0), \tag{6.41}$$

$$W_{\mathbf{A}_j,\blacklozenge,\text{avoided}}(t_0) \supseteq \underbrace{W_{\mathbf{B}_i,\blacklozenge}(t_0) \supseteq W_{\mathbf{A}_i,\blacklozenge}(t_0)}_{\text{from lines 8 and 30 of Algorithm 6.1}}, \tag{6.42}$$

which consequently leads to

$$\underbrace{W_{\mathbf{A}_j,\blacklozenge}(t_0) \cap \left(W_{\mathbf{B}_i,\blacklozenge}(t_0) \setminus W_{\mathbf{A}_i,\blacklozenge,\text{avoided}}(t_0)\right)}_{\text{since (6.41)}} = \varnothing, \tag{6.43}$$

$$\underbrace{W_{\mathbf{A}_i,\blacklozenge}(t_0) \cap \left(W_{\mathbf{B}_j,\blacklozenge}(t_0) \setminus W_{\mathbf{A}_j,\blacklozenge,\text{avoided}}(t_0)\right)}_{\text{since (6.42)}} = \varnothing, \tag{6.44}$$

and

$$\underbrace{\left(W_{\mathbf{B}_i,\blacklozenge}(t_0) \setminus W_{\mathbf{A}_i,\blacklozenge,\text{avoided}}(t_0)\right) \cap \left(W_{\mathbf{B}_j,\blacklozenge}(t_0) \setminus W_{\mathbf{A}_j,\blacklozenge,\text{avoided}}(t_0)\right)}_{\text{since (6.42)}} = \varnothing. \tag{6.45}$$

By substituting (6.40) into (6.39) and from (6.43), (6.44) and (6.45), one obtains

$$\begin{aligned}
T_5 = {}& \underbrace{\left[W_{\mathbf{A}_i,\blacklozenge}(t_0) \cap W_{\mathbf{A}_j,\blacklozenge}(t_0)\right]}_{=\varnothing,\,\text{from (6.38)}} \cap \underbrace{\left[W_{\mathbf{A}_i,\blacklozenge}(t_0) \cap \left(W_{\mathbf{B}_j,\blacklozenge}(t_0) \setminus W_{\mathbf{A}_j,\blacklozenge,\text{avoided}}(t_0)\right)\right]}_{=\varnothing,\,\text{from (6.44)}} \\
& \cap \underbrace{\left[W_{\mathbf{A}_j,\blacklozenge}(t_3) \cap \left(W_{\mathbf{B}_i,\blacklozenge}(t_0) \setminus W_{\mathbf{A}_i,\blacklozenge,\text{avoided}}(t_0)\right)\right]}_{=\varnothing,\,\text{see (6.43)}} \\
& \cap \underbrace{\left[\left(W_{\mathbf{B}_i,\blacklozenge}(t_0) \setminus W_{\mathbf{A}_i,\blacklozenge,\text{avoided}}(t_0)\right) \cap \left(W_{\mathbf{B}_j,\blacklozenge}(t_0) \setminus W_{\mathbf{A}_j,\blacklozenge,\text{avoided}}(t_0)\right)\right]}_{=\varnothing,\,\text{from (6.45)}} \\
= {}& \varnothing, \quad \forall\, t_0 < t \le t_2.
\end{aligned} \tag{6.46}$$

- $T_6 = T_8 = \varnothing$:

In analogy to T_3 and T_7, T_6 and T_8 are also symmetrical by interchanging the agent and robot indexes. Consequently, it is sufficient to show that $T_6 = \varnothing$. In fact, one has from (6.11), (6.26) and (6.37) that

$$T_6 = W_{\mathbf{A}_{i,\blacklozenge}}(t) \cap \mathbf{A}_j^{\star}(t), \tag{6.47}$$

$$\mathbf{A}_i^{\star}(t) \subseteq W_{\mathbf{A}_i,\star}(t) \tag{6.48}$$

$$W_{\mathbf{A}_{i,\blacklozenge}}(t_0) \cap W_{\mathbf{A}_{j,\star}}(t) = \varnothing, \quad \forall\, t, \quad i \ne j, \quad \forall\, \blacklozenge, \star. \tag{6.49}$$

By substituting (6.48) and (6.49) into (6.47) and from (6.36), one has

$$T_6 = W_{\mathbf{A}_{i,\blacklozenge}}(t) \cap \mathbf{A}_j^{\star}(t) \subseteq W_{\mathbf{A}_{i,\blacklozenge}}(t) \cap W_{\mathbf{A}_{j,\star}}(t) = \varnothing \quad \forall t. \tag{6.50}$$

- $T_9 = \varnothing$:

From from (6.11), one has that

$$T_9 = \mathbf{A}_i^{\clubsuit}(t) \cap \mathbf{A}_j^{\star}(t). \tag{6.51}$$

From (6.26) and (6.36), one has that

$$\begin{aligned}
\mathbf{A}_i^{\clubsuit}(t) &\subseteq W_{\mathbf{A}_i,\clubsuit}^{\clubsuit}(t), \\
\mathbf{A}_j^{\star}(t) &\subseteq W_{\mathbf{A}_j,\star}^{\star}(t), \\
W_{\mathbf{A}_{i,\clubsuit}}^{\clubsuit}(t) \cap W_{\mathbf{A}_{j,\star}}^{\star}(t) &= \varnothing, \quad \forall\, t, \quad i \neq j, \quad \forall\, \clubsuit, \star.
\end{aligned} \tag{6.52}$$

From (6.36), substituting (6.52) into (6.51) leads to

$$T_9 = \mathbf{A}_i^{\clubsuit}(t) \cap \mathbf{A}_j^{\star}(t) \subseteq W_{\mathbf{A}_{i,\clubsuit}}^{\clubsuit}(t) \cap W_{\mathbf{A}_{j,\star}}^{\star}(t) = \varnothing, \quad \forall t. \tag{6.53}$$

In conclusion, it has been shown that $T_i = \varnothing$, $\forall i \in \{1, \ldots, 9\}$. Therefore, we have from (6.4) that

$$W_{\mathbf{A}_i}^{\blacklozenge}(t) \cap W_{\mathbf{A}_j}^{\blacklozenge}(t) = \varnothing, \quad \forall i \neq j, \quad i \in W_{\mathcal{M}_{\mathrm{on}}}^{\blacklozenge} \wedge j \in W_{\mathcal{M}_{\mathrm{on}}}^{\blacklozenge}, \tag{6.54}$$

which implies that the distributed collision avoidance algorithm is invariantly safe ■.

6.7 Computation Complexity and Simulations

In this section, the computation complexity of the distributed collision avoidance algorithm is analyzed. As mentioned in Section 6.2, the distributed collision avoidance algorithm is comprised of the state transition, the trajectory generation and the message generation functions. Therefore, the computational complexity of these functions is subsequently investigated. Let $N_{\mathrm{r,on}}$ denote the number of robots placed on an agent, while N_r is the defined as the number of robots related on an agent. Then, the computational complexity is determined in the following.

In the trajectory optimization presented in Chapter 4, the maximal number of the geometrical way points can be specified. Therefore, the reserved and the requested areas can be represented by a limited number of convex hulls. As a consequence, the computational complexity of the basic geometrical operations on these areas is $\mathcal{O}(1)$. Moreover, these operations are used from line 1 to 22 of Algorithm 6.1, which leads to a computational burden of $\mathcal{O}(1)$. Furthermore, the computational complexity is proportional to $N_{\mathrm{r,on}}$ for lines 23-35 and 47-54 and quadratic to N_r for lines 36-36. Therefore, the state transition function shown in Algorithm 6.1 has the computational complexity of $\mathcal{O}(N_{\mathrm{r,on}} + N_\mathrm{r}^2)$. From Algorithm 6.2, the computation complexity of the trajectory generation function can be defined, which is proportional to $N_{\mathrm{r,on}}$ as $\mathcal{O}(N_{\mathrm{r,on}})$. Similarly, the computational complexity of the message generation function can be determined as $\mathcal{O}(N_\mathrm{r})$ from Algorithm 6.3. Eventually, the computational complexity of the asynchronous distributed collision avoidance algorithm can be aggregated as

$$\mathcal{O}(N_{\mathrm{r,on}} + N_\mathrm{r}^2) + \mathcal{O}(N_{\mathrm{r,on}}) + \mathcal{O}(N_\mathrm{r}) = \mathcal{O}(N_{\mathrm{r,on}} + N_\mathrm{r}^2).$$

Moreover, one has that $N_{r,on} \leq N_r$, since $W_{\mathcal{M}_{on}}^{\blacklozenge} \subseteq W_{\mathcal{M}}^{\blacklozenge}$ as mentioned in Section 6.2. Therefore, this computational complexity is equivalent to $\mathcal{O}(N_r^2)$. In other words, the computational complexity of the distributed collision avoidance algorithm increased quadratically with the number of robots related to the agent.

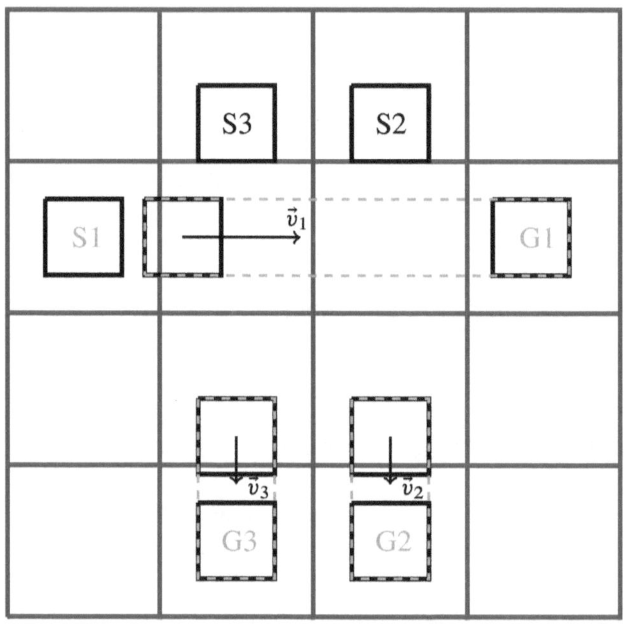

Figure 6.7: The simulation framework for verifying the distributed collision avoidance algorithm

In order to validate the invariant safety of the distributed collision avoidance algorithm, a framework is simulated in MATLAB® 2017b environment. As shown in Figure 6.7, the simulation framework consists of sixteen rectangle agents and three rectangle robots. The width of each agents and robots are respectively $0.24\ m$ and $0.12\ m$. The stochastic communication time delays among agents are uniformly distributed between $0.005s$ and $0.025s$. The robots are assigned between start and goal locations and back again. The robot kinematic constraints are identical, which are respectively $\mathbf{v}_{max} = [3 \quad 3]^T \frac{m}{s}$, $\mathbf{a}_{max} = [20 \quad 20]^T \frac{m}{s^2}$ and $\mathbf{j}_{max} = [800 \quad 800]^T \frac{m}{s^3}$. As theoretically expected, no collision is found during the simulation. Furthermore, the robot reserved areas $W_{\mathbf{A}_i}^{\blacklozenge}$, $i \in \{1, 2, 3\}$, are also visualized during the simulation. No overlapping among these assigned working areas is observed, which confirms Theorem 6.1.

6.8 Conclusion

In this chapter, a novel distributed collision avoidance algorithm is presented, which is robust against stochastic communication time delays. Particularly, the algorithm is based on the reserved and requested area concept. By utilizing the convex hull property of the generated trajectories, these reserved and requested areas can be computed and updated efficiently. Furthermore, the invariant safety of the algorithm is formally proved and validated in simulations. Additionally, the computational complexity of the algorithm is also analyzed, which increases quadratically to the number of robots related to the deployed agent. In order to accomplish the solution, the future work is devoted to develop a distributed conflict resolution algorithm.

7 Conclusion

This chapter is intended to review and collect all novel results, which are presented throughout the book. Furthermore, their applications to real industrial systems and extensions for future works are discussed.

Firstly, a novel high-order near time-optimal trajectory generation algorithm is presented in Chapter 4, which is not only numerical efficient but also able to take all high-order time derivative constraints into account. Moreover, the high-order time derivatives are only discontinuous at time stamps t_{d}, which are integer numbers of the sampling time T_{s}. This essential time discretization is considered for high-performance motion control applications in semiconductor industry [8]. Additionally, a termination analysis of the algorithm is also provided. Eventually, a short comparison with the state of the art approaches is given. Therefore, the primary goal for future works is applying the algorithm to existing industrial systems in order to make an impact.

Subsequently, a highly modular novel centralized coordination solution is introduced in Chapter 5, which consists of a collision avoidance algorithm and a conflict resolution algorithm. In particular, the collision avoidance algorithm is able to guarantee safety invariantly, while the conflict resolution algorithm is employed to prevent deadlocks. Moreover, the invariant safety of the centralized collision avoidance algorithm can be formally proved and verified in several simulations. Furthermore, the centralized coordination solution can be applied as a post process to the proposed trajectory generation algorithm. Consequently, the high-order time derivative constraints as well as the time discretization effect consideration can be preserved. According to our knowledge, no existing centralized collision avoidance algorithm is able to take the high-order time derivative constraints and the time discretization into account, which simultaneously has a small computational burden. As a consequence, one is able to integrate these algorithms directly into the commercial high performance motion control systems [8] in order to avoid collisions.

Finally, an asynchronous distributed collision avoidance algorithm is presented in Chapter 6, which is developed for a group of robots translating on a grid of rectangular intelligent agents. Furthermore, the algorithm is not only able to guarantee collision-free motions but also robust with respect to the stochastic arbitrarily large communication time delays. The invariant safety guarantee of the algorithm is also formally proved. Additionally, the algorithm is based on the reserved and requested areas, which can be efficiently computed by utilizing the convex hull property of the generated trajectories. Therefore, it can be seamlessly integrated with the trajectory generation algorithm. Nevertheless, the state of the art distributed conflict resolution algorithm can only be applied for primitive four cardinal motions [34]. Therefore, there is a substantial motivation to develop a novel distributed conflict resolution algorithm in order to utilize the flexible motions. Together with the asynchronous distributed collision avoidance algorithm, it will scale existing industrial motion control systems to another dimension.

95

Appendix

A.1 A Formal Proof of Theorem 4.1

Theorem A.1 *Suppose that Assumption 4.1 holds and $j_{max} > 0$, then the condition $\max |\dddot{q}_\xi(t)| \leq j_{max}$ will be satisfied after a finite number of repeat loops by using Algorithm 4.1.*

Proof Assume that the condition $\max \dddot{q}_\xi(t)$ is not satisfied as the number of repeat loop tends to infinity. There is at least a spline segment k such that

$$\max |\dddot{S}_{k,\xi}(t)| > j_{max},$$

for arbitrarily large h_k. Moreover, we have from (4.13) that

$$\max |\dddot{S}_{k,\xi}(t)| = \left| \frac{M_{k+1,\xi} - M_{k,\xi}}{h_k} \right| \tag{A.1}$$

where $M_\xi = A^{-1}\mathbf{d}_\xi$. Let denote $a_{k,j}^{-1}$ is the element of A^{-1}. Thus, one has

$$M_{k,\xi} = \sum_{j=0}^{N_s} a_{k,j}^{-1} d_{j,\xi}, \quad M_{k+1,\xi} = \sum_{j=0}^{N_s} a_{k+1,j}^{-1} d_{j,\xi}. \tag{A.2}$$

Moreover, the matrix A is a strictly diagonal dominant. By using the results in [103], one consequently obtains

$$\|A^{-1}\|_\infty \leq \max_{i=\{0,1,\dots,N_s\}} \left\{ \frac{1}{|a_{ii}| - \sum\limits_{j=0,j\neq i}^{N_s} |a_{ij}|} \right\} = \max_{i=\{0,1,\dots,N_s\}} \frac{2}{2 - \mu_i - (1 - \mu_i)} = 1 \tag{A.3}$$

$$\Rightarrow \sum_{j=0}^{N_s} |a_{k,j}^{-1}| \leq 1 \quad \text{and} \quad \sum_{j=0}^{N_s} |a_{k+1,j}^{-1}| \leq 1.$$

Substituting (A.2) and (A.3) into (A.1) yields

$$\lim_{h_k \to \infty} |\dddot{S}_{k,\xi}(t)| = \lim_{h_k \to \infty} \left| \frac{M_{k+1,\xi} - M_{k,\xi}}{h_k} \right| = \lim_{h_k \to \infty} \left| \frac{\sum\limits_{j=0}^{N_s} a_{k+1,j} d_{j,\xi} - \sum\limits_{j=0}^{N_s} a_{k+1,j} d_{j,\xi}}{h_k} \right|$$

$$\leq \lim_{h_k \to \infty} \frac{\sum\limits_{j=0}^{N_s} |a_{k+1,j}||d_{j,\xi}| + \sum\limits_{j=0}^{N_s} |a_{k+1,j}||d_{j,\xi}|}{h_k}$$

$$\leq \lim_{h_k \to \infty} \underbrace{\frac{2 \max\limits_{j=0,1,\ldots,N_s} |d_{j,\xi}|}{h_k}}_{\max\limits_{j=0,1,\ldots,N_s} |d_{j,\xi}| \text{ is limited.}} = 0 \quad \geq \quad j_{\max},$$

which is a contradiction to $j_{\max} > 0$. Therefore, the condition $\max |\dddot{q}_\xi(t)| \leq j_{\max}$ is satisfied after a finite number of repeat loops of Algorithm 4.1 ($\blacksquare$).

Theorem A.2 *Suppose that Assumption 4.1 holds and $a_{max} > 0$. Then, the condition $\max |\ddot{q}_\xi(t)| \leq a_{max}$, $\forall \xi \in \{x, y\}$, will be satisfied after a finite number of repeat loops by means of Algorithm 4.1.*

Proof Assume that the condition $\max |\ddot{q}_\xi(t)| \leq a_{\max}$ is not satisfied, as the number of repeat loop tends to infinity. There is at least a spline segment k such that

$$\max |\ddot{S}_{k,\xi}(t)| > a_{\max}, \tag{A.5}$$

for arbitrarily large h_k. Let $S_{k_{\max},\xi}$ be the segment with the maximal absolute acceleration value of the coordinates $q_\xi(t)$. Then, the maximal absolute value of acceleration of the coordinate ξ on the segment $S_{k_{\max},\xi}$ can be defined as

$$|\ddot{S}_{k_{\max},\xi}| = \max\{|M_{k_{\max},\xi}|, |M_{k_{\max}+1,\xi}|\}.$$

Furthermore, one obtains from (4.11) that

$$\begin{bmatrix} 2 & 1 - \mu_{k_{\max}} \\ \mu_{k_{\max}+1} & 2 \end{bmatrix} \begin{bmatrix} M_{k_{\max},\xi} \\ M_{k_{\max}+1,\xi} \end{bmatrix} = \begin{bmatrix} d_{k_{\max},\xi} - \mu_{k_{\max}} M_{k_{\max}-1,\xi} \\ d_{k_{\max}+1} - (1 - \mu_{k_{\max}+1}) M_{k_{\max}+2,\xi} \end{bmatrix}. \tag{A.6}$$

Solving the system of two equations (A.6), we have that

$$\begin{aligned} M_{k_{max},\xi} = & \frac{2}{4 - (1 - \mu_{k_{\max}})\mu_{k_{\max}+1}} (d_{k_{\max},\xi} - \mu_{k_{\max}} M_{k_{\max}-1,\xi}) \\ & - \frac{1 - \mu_{k_{\max}}}{4 - (1 - \mu_{k_{\max}})\mu_{k_{\max}+1}} \left(d_{k_{\max}+1} - (1 - \mu_{k_{\max}+1}) M_{k_{\max}+2} \right), \end{aligned} \tag{A.7}$$

and

$$\begin{aligned} M_{k_{max}+1,\xi} = & -\frac{\mu_{k_{\max}+1}}{4 - (1 - \mu_{k_{\max}})\mu_{k_{\max}+1}} (d_{k_{\max},\xi} - \mu_{k_{\max}} M_{k_{\max}-1,\xi}) \\ & + \frac{2}{4 - (1 - \mu_{k_{\max}})\mu_{k_{\max}+1}} \left(d_{k_{\max}+1,\xi} - (1 - \mu_{k_{\max}+1}) M_{k_{\max},\xi+2,\xi} \right). \end{aligned} \tag{A.8}$$

Now, assume that

$$|\ddot{S}_{k_{\max},\xi}| = |M_{k_{\max},\xi}|.$$

The same logic can be applied for another case $|\ddot{S}_{k_{\max},\xi}| = |M_{k_{\max}+1,\xi}|$. From (A.7), we have that

$$
\begin{aligned}
|M_{k_{max},\xi}| &\leq \frac{2|d_{k_{\max},\xi}| + |1 - \mu_{k_{\max}}||d_{k_{\max}+1,\xi}|}{|4 - (1 - \mu_{k_{\max}})\mu_{k_{\max}+1}|} \\
&\quad + \frac{2|\mu_{k_{\max}} M_{k_{\max}-1,\xi}| + |(1 - \mu_{k_{\max}})(1 - \mu_{k_{\max}+1})M_{k_{\max}+2,\xi}|}{|4 - (1 - \mu_{k_{\max}})\mu_{k_{\max}+1}|} \\
&\leq \frac{2|d_{k_{\max},\xi}| + |1 - \mu_{k_{\max}}||d_{k_{\max}+1,\xi}|}{|4 - (1 - \mu_{k_{\max}})\mu_{k_{\max}+1}|} \\
&\quad + \underbrace{\frac{2|\mu_{k_{\max}}| + |(1 - \mu_{k_{\max}})(1 - \mu_{k_{\max}+1})|}{|4 - (1 - \mu_{k_{\max}})\mu_{k_{\max}+1}|}}_{\leq \frac{1}{2},\ \text{since}\quad 0 \leq \mu_{k_{\max}}, \mu_{k_{\max}+1} \leq 1}\ \underbrace{|M_{k_{\max},\xi}|}_{\geq |M_{k_{\max}-1,\xi}|, |M_{k_{\max}+2,\xi}|} \\
&\leq \frac{2|d_{k_{\max},\xi}| + |1 - \mu_{k_{\max}}||d_{k_{\max}+1,\xi}|}{|4 - (1 - \mu_{k_{\max}})\mu_{k_{\max}+1}|} + \frac{1}{2}|M_{k_{\max},\xi}| \\
\Rightarrow |M_{k_{max},\xi}| &\leq 2\frac{2|d_{k_{\max},\xi}| + |1 - \mu_{k_{\max}}||d_{k_{\max}+1,\xi}|}{|4 - (1 - \mu_{k_{\max}})\mu_{k_{\max}+1}|}.
\end{aligned}
\tag{A.9}
$$

From (4.11), (4.12) and (A.9), one has

$$
\begin{aligned}
&\lim_{h_{k_{\max}} \to \infty} d_{k_{\max},\xi} = \lim_{h_{k_{\max}} \to \infty} d_{k_{\max}+1,\xi} = 0, \\
&\Rightarrow \lim_{h_{k_{\max}} \to \infty} |\ddot{S}_{k_{\max},\xi}(t)| = \lim_{h_{k_{\max}} \to \infty} |M_{k_{\max},\xi}(t)| = 0
\end{aligned}
\tag{A.10}
$$

Thus, we have from (A.5) and (A.10) that

$$
a_{\max} \leq \inf_{h_{k_{\max}} > 0} |\ddot{S}_{k_{\max},\xi}(t)| = \lim_{h_{k_{\max}} \to \infty} |\ddot{S}_{k_{\max},\xi}(t)| = 0,
\tag{A.11}
$$

which obviously leads to a contradiction ($\blacksquare$).

Lemma A.1 *Suppose that Assumption 4.1 is satisfied. Then,*

$$
\lim_{h_k \to \infty} \max_{t_{k-1} \leq t \leq t_k} |\dot{S}_{k,\xi}(t)| = \max \left\{ \lim_{h_k \to \infty} |\dot{S}_{k,\xi}(t = t_{k-1})|,\ \lim_{h_k \to \infty} |\dot{S}_{k,\xi}(t = t_k)| \right\}.
$$

Proof From (4.8), one has

$$
\max_{t_{k-1} \leq t \leq t_k} |\dot{S}_{k,\xi}(t)| = \max \left\{ |\dot{S}_{k,\xi}(t = t_{k-1})|, |\dot{S}_{k,\xi}(t = t_k)|, |\dot{S}_{k,\xi}(t = \tau)| \right\},
$$

where

$$
|\dot{S}_{k,\xi}(t = t_{k-1})| = \left| \frac{q_{k,\xi} - q_{k-1,\xi}}{h_k} - \frac{(2M_{k-1,\xi} + M_{k,\xi})h_k}{6} \right|,
\tag{A.12}
$$

$$
|\dot{S}_{k,\xi}(t = t_k)| = \left| \frac{q_{k,\xi} - q_{k-1,\xi}}{h_k} + \frac{(M_{k-1,\xi} + 2M_{k,\xi})h_k}{6} \right|,
\tag{A.13}
$$

and

$$|\dot{S}_{k,\xi}(t=\tau)| = \begin{cases} \left| \frac{q_{k,\xi}-q_{k-1,\xi}}{h_k} + \frac{(M_{k-1,\xi}M_{k,\xi})h_k}{2(M_{k-1,\xi}-M_{k,\xi})} + \frac{(M_{k-1,\xi}-M_{k,\xi})h_k}{6} \right|, & \text{if } 0 < \frac{M_{k-1}}{M_{k-1}-M_k} < 1, \\ 0, & \text{otherwise.} \end{cases}$$

(A.14)

Applying the maximum limitation law [99], we obtain

$$\lim_{h_k \to \infty} \max_{t_{k-1} \le t \le t_k} |\dot{S}_{k,\xi}(t)| = \max\left\{ \underbrace{\lim_{h_k \to \infty} |\dot{S}_{k,\xi}(t=t_{k-1})|}_{L_1}, \underbrace{\lim_{h_k \to \infty} |\dot{S}_{k,\xi}(t=t_k)|}_{L_2}, \underbrace{\lim_{h_k \to \infty} |\dot{S}_{k,\xi}(t=\tau)|}_{L_3} \right\} = \max\{L_1, L_2, L_3\}.$$

(A.15)

Furthermore, one has from (A.12), (A.13) and (A.14) that

$$L_1 = \lim_{h_k \to \infty} \left| \frac{(2M_{k-1,\xi} + M_{k,\xi})h_k}{6} \right|,$$

(A.16)

$$L_2 = \lim_{h_k \to \infty} \left| \frac{(M_{k-1,\xi} + 2M_{k,\xi})h_k}{6} \right| \text{ and}$$

(A.17)

$$L_3 = \lim_{h_k \to \infty} \left| \frac{(M_{k-1,\xi}M_{k,\xi})h_k}{2(M_{k-1,\xi} - M_{k,\xi})} + \frac{(M_{k-1,\xi} - M_{k,\xi})h_k}{6} \right| < \max\{L_1, L_2\},$$

(A.18)

since

$$\left| \frac{(M_{k-1,\xi}M_{k,\xi})h_k}{2(M_{k-1,\xi}-M_{k,\xi})} + \frac{(M_{k-1,\xi}-M_{k,\xi})h_k}{6} \right| < \max\left\{ \left| \frac{(2M_{k-1,\xi}+M_{k,\xi})h_k}{6} \right|, \left| \frac{(M_{k-1,\xi}+2M_{k,\xi})h_k}{6} \right| \right\}$$

for $0 < \frac{M_{k-1,\xi}}{M_{k-1,\xi}-M_{k,\xi}} < 1$. From (A.15), (A.16), (A.17) and (A.18), one has

$$\lim_{h_k \to \infty} \max_{t_{k-1} \le t \le t_k} |\dot{S}_{k,\xi}(t)| = \max\{L_1, L_2\}$$

$$= \max\left\{ \lim_{h_k \to \infty} |\dot{S}_{k,\xi}(t=t_{k-1})|, \lim_{h_k \to \infty} |\dot{S}_{k,\xi}(t=t_k)| \right\} \; (\blacksquare).$$

Furthermore, the strict inequality (A.18) implies that there exists a h_k^* such that

$$\max_{t_{k-1} \le t \le t_k} |\dot{S}_{k,\xi}(t)| = \max\left\{ |\dot{S}_{k,\xi}(t=t_{k-1})|, |\dot{S}_{k,\xi}(t=t_k)| \right\}, \; \forall\, h_k \ge h_k^*,$$

or

$$\lim_{h_k \to \infty} \max_{t_{k-1} \le t \le t_k} |\dot{S}_{k,\xi}(t)| = 0.$$

In order to simplify the notation, the coordinate ξ is omitted in the following. By solving the following equation system

$$\begin{bmatrix} 2 & 1-\mu_{N-1} \\ \mu_N & 2 \end{bmatrix} \begin{bmatrix} M_{N-1} \\ M_N \end{bmatrix} = \begin{bmatrix} d_{N-1} - \mu_{N-1}M_{N-2} \\ d_N \end{bmatrix},$$

(A.19)

one obtains that

$$M_{N-1} = \frac{2}{4 - (1 - \mu_{N-1})\mu_N}(d_{N-1} - \mu_{N-1}M_{N-2}) - \frac{(1 - \mu_{N-1})}{4 - (1 - \mu_{N-1})\mu_N}d_N, \qquad \text{(A.20)}$$

$$M_N = -\frac{\mu_N}{4 - (1 - \mu_{N-1})\mu_N}(d_{N-1} - \mu_{N-1}M_{N-2}) + \frac{2}{4 - (1 - \mu_{N-1})\mu_N}d_N. \qquad \text{(A.21)}$$

Replacing (A.21) in into (A.19), we have

$$\underbrace{\frac{4\mu_{N-1}}{4 - (1 - \mu_{N-1})\mu_N}}_{=\tilde{\mu}_{N-1}<1, \quad \text{since}\,\mu_N \leq 1} M_{N-2} + 2M_{N-1} = \underbrace{\frac{4}{4 - (1 - \mu_{N-1})\mu_N}d_{N-1} - \frac{2(1 - \mu_{N-1})}{4 - (1 - \mu_{N-1})\mu_N}d_N}_{\tilde{d}_{N-1}}$$

Analogously, we have that

$$\underbrace{\frac{4\mu_{N-2}}{4 - (1 - \mu_{N-2})\tilde{\mu}_{N-1}}}_{=\tilde{\mu}_{N-2}<1, \quad \text{since}\,\tilde{\mu}_{N-1}\leq 1} M_{N-3} + 2M_{N-2} = \underbrace{\frac{4}{4 - (1 - \mu_{N-2})\tilde{\mu}_{N-1}}d_{N-2} - \frac{2(1 - \mu_{N-2})}{4 - (1 - \mu_{N-2})\tilde{\mu}_{N-1}}\tilde{d}_{N-1}}_{=\tilde{d}_{N-2}},$$

$$\vdots \qquad\qquad\qquad\qquad \vdots$$

$$\underbrace{\frac{4\mu_{k+2}}{4 - (1 - \mu_{k+2})\tilde{\mu}_{k+3}}}_{=\tilde{\mu}_{k+2}<1} M_{k+1} + 2M_{k+2} = \underbrace{\frac{4}{4 - (1 - \mu_{k+2})\tilde{\mu}_{k+3}}d_{k+2} - \frac{2(1 - \mu_{k+2})}{4 - (1 - \mu_{k+2})\tilde{\mu}_{k+3}}\tilde{d}_{k+3}}_{=\tilde{d}_{k+2}},$$

$$\text{(A.22)}$$

By substituting (A.22) into $\tilde{d}_{k+2}h_{k+2}$, one has

$$
\begin{aligned}
\frac{1}{6}\tilde{d}_{k+2}h_{k+2} &= \frac{4}{4 - (1 - \mu_{k+2})\tilde{\mu}_{k+3}}\frac{d_{k+2}h_{k+2}}{6} - \frac{2(1 - \mu_{k+2})}{4 - (1 - \mu_{k+2})\tilde{\mu}_{k+3}}\frac{\tilde{d}_{k+3}h_{k+2}}{6} \\
&= \frac{4}{4 - (1 - \mu_{k+2})\tilde{\mu}_{k+3}}\frac{d_{k+2}h_{k+2}}{6} - \frac{2\mu_{k+2}}{4 - (1 - \mu_{k+2})\tilde{\mu}_{k+3}}\frac{\tilde{d}_{k+3}h_{k+3}}{6} \\
&= \underbrace{\frac{4\mu_{k+2}}{4 - (1 - \mu_{k+2})\tilde{\mu}_{k+3}}}_{0\leq\gamma_{k+2}\leq 1}\left(\frac{q_{k+3} - q_{k+2}}{h_{k+3}} - \frac{q_{k+2} - q_{k+1}}{h_{k+2}}\right) \\
&\quad - \underbrace{\frac{2\mu_{k+2}}{4 - (1 - \mu_{k+2})\tilde{\mu}_{k+3}}}_{\frac{1}{2}\gamma_{k+2}}\frac{\tilde{d}_{k+3}h_{k+3}}{6} \\
&= -\gamma_{k+2}\frac{q_{k+2} - q_{k+1}}{h_{k+2}} + \gamma_{k+2}\frac{q_{k+3} - q_{k+2}}{h_{k+3}} \\
&\quad - \frac{1}{2}\gamma_{k+2}\left(\gamma_{k+3}(\frac{q_{k+4} - q_{k+3}}{h_{k+4}} - \frac{q_{k+3} - q_{k+2}}{h_{k+3}}) - \frac{1}{2}\gamma_{k+3}\frac{\tilde{d}_{k+4}h_{k+4}}{6}\right) \\
&= \gamma_{k+2}\frac{q_{k+1} - q_{k+2}}{h_{k+2}} - \gamma_{k+2}(1 + \frac{\gamma_{k+3}}{2})\frac{q_{k+2} - q_{k+3}}{h_{k+3}} \\
&\quad + \frac{1}{2}\gamma_{k+2}\gamma_{k+3}(1 + \frac{\gamma_{k+4}}{2})\frac{q_{k+3} - q_{k+4}}{h_{k+4}} + \cdots
\end{aligned}
\qquad \text{(A.23)}
$$

$$+ \frac{(-1)^{N-k-2}}{2^{N-k-3}} \gamma_{k+2}\gamma_{k+3}\cdots\gamma_{N-1}(1 + \underbrace{\gamma_N}_{=\frac{1}{2}})\frac{q_{N-1} - q_N}{h_N}$$

By solving the following equation system

$$\begin{bmatrix} 2 & \mu_0 \\ \mu_1 & 2 \end{bmatrix}\begin{bmatrix} M_0 \\ M_1 \end{bmatrix} = \begin{bmatrix} d_0 \\ d_1 - (1 - \mu_1)M_2 \end{bmatrix}, \tag{A.24}$$

one obtains that

$$M_0 = \frac{2}{4 - \mu_0\mu_1}d_0 - \frac{\mu_0}{4 - \mu_0\mu_1}(d_1 - (1 - \mu_1)M_2) \tag{A.25}$$

Replacing (A.25) into (A.24), one has that

$$2M_1 + (1 - \mu_1)M_2 = d_1 - \mu_1 M_0 = \frac{4}{4 - \mu_0\mu_1}d_1 - \frac{2\mu_1}{4 - \mu_0\mu_1}d_0 - \frac{\mu_0\mu_1(1 - \mu_1)}{4 - \mu_0\mu_1}M_2$$

$$\Leftrightarrow 2M_1 + \underbrace{\frac{4(1 - \mu_1)}{4 - \mu_0\mu_1}}_{=(1-\tilde{\mu}_1)<1} M_2 = \underbrace{\frac{4}{4 - \mu_0\mu_1}d_1 - \frac{2\mu_1}{4 - \mu_0\mu_1}d_0}_{\tilde{d}_1}.$$

Analogously, we have

$$2M_2 + \underbrace{\frac{4(1 - \mu_2)}{4 - \tilde{\mu}_1\mu_2}}_{=(1-\tilde{\mu}_2)<1} M_3 \quad = \quad \underbrace{\frac{4}{4 - \tilde{\mu}_1\mu_2}d_2 - \frac{2\mu_2}{4 - \tilde{\mu}_1\mu_2}\tilde{d}_1}_{\tilde{d}_2}.$$

$$\vdots \qquad\qquad\qquad \vdots$$

$$2M_{k-2} + \underbrace{\frac{4(1 - \mu_{k-2})}{4 - \tilde{\mu}_{k-3}\mu_k}}_{=(1-\tilde{\mu}_{k-2})<1} M_{k-1} \quad = \quad \underbrace{\frac{4}{4 - \tilde{\mu}_{k-3}\mu_{k-2}}d_{k-2} - \frac{2\mu_{k-2}}{4 - \tilde{\mu}_{k-3}\mu_{k-2}}\tilde{d}_{k-3}}_{\tilde{d}_{k-2}}.$$

Therefore, we have

$$\frac{1}{6}\tilde{d}_{k-2}h_{k-1} = \underbrace{\frac{4(1 - \mu_{k-2})}{4 - \tilde{\mu}_{k-3}\mu_{k-2}}}_{\beta_{k-2}}\left(\frac{q_{k-1} - q_{k-2}}{h_{k-1}} - \frac{q_{k-2} - q_{k-3}}{h_{k-2}}\right) - \underbrace{\frac{2(1 - \mu_{k-2})}{4 - \tilde{\mu}_{k-3}\mu_{k-2}}}_{\frac{1}{2}\beta_{k-2}}\frac{\tilde{d}_{k+3}h_{k-2}}{6}$$

$$= \beta_{k-2}\frac{q_{k-1} - q_{k-2}}{h_{k-1}} - \beta_{k-2}(1 + \frac{\beta_{k-3}}{2})\frac{q_{k-2} - q_{k-3}}{h_{k-2}}$$

$$+ \frac{1}{2}\beta_{k-2}\beta_{k-3}(1 + \frac{\beta_{k-4}}{2})\frac{q_{k-3} - q_{k-4}}{h_{k-3}} + \dots.$$

$$+ \frac{(-1)^{k-2}}{2^{k-3}}\beta_{k-2}\beta_{k-3}..\beta_1(\frac{q_1 - q_0}{h_1} + \frac{1}{12}d_0h_1)$$

$$\tag{A.26}$$

Lemma A.2 *Suppose that Assumptions 4.1 and 4.2 are satisfied. There exists a $\delta > 0$ such that*

$$\lim_{\substack{h_{k+1}\to\infty \\ h_k\to\infty}} |\dot{S}_{k,\xi}(t = t_k)| \leq v_{max} - \delta, \ \forall\, k = \{0, 1, \dots, N_s\}$$

by means of Algorithm 4.1.

Proof The proof of the inequality

$$\lim_{\substack{h_{k+1}\to\infty \\ h_k\to\infty}} |\dot{S}_{k,\xi}(t = t_k)| \le v_{\max} - \delta$$

is sketched as follows. Without loss of generality, assume that $h_{k+1} = r h_k + z$, where $r \in \mathbb{R}^+$ and $z \in \mathbb{Z}$ is a finite number. By solving the following equation system

$$\begin{bmatrix} 2 & 1-\tilde{\mu}_{k-2} & 0 & 0 & 0 \\ \mu_{k-1} & 2 & 1-\mu_{k-1} & 0 & 0 \\ 0 & \mu_k & 2 & 1-\mu_k & 0 \\ 0 & 0 & \mu_{k+1} & 2 & 1-\mu_{k+1} \\ 0 & 0 & 0 & \tilde{\mu}_{k+2} & 2 \end{bmatrix} \begin{bmatrix} M_{k-2} \\ M_{k-1} \\ M_k \\ M_{k+1} \\ M_{k+2} \end{bmatrix} = \begin{bmatrix} \tilde{d}_{k-2} \\ d_{k-1} \\ d_k \\ d_{k+1} \\ \tilde{d}_{k+2}. \end{bmatrix}$$

and substituting the solution into (A.13), one obtains

$$\lim_{\substack{h_{k+1}\to\infty \\ h_k\to\infty}} |\dot{S}_{k,\xi}(t = t_k)| \le \left| \frac{r}{r+1}\left(\frac{q_{k-2,\xi} - q_{k-1,\xi}}{2h_{k-1}} - \frac{\tilde{d}_{k-2,\xi}h_{k-1}}{24} \right) \right| \tag{A.27}$$
$$+ \left| \frac{1}{r+1}\left(\frac{q_{k+1,\xi} - q_{k+2,\xi}}{2h_{k+2}} + \frac{\tilde{d}_{k+2,\xi}h_{k+2}}{24} \right) \right|$$

By substituting the above derived recursive formulas (A.23) and (A.26) of the terms $\tilde{d}_{k+2,\xi}h_{k+2}$, $\tilde{d}_{k-2,\xi}h_{k-1}$ into (A.27) and using Assumption 4.2, we have

$$\lim_{\substack{h_{k+1}\to\infty \\ h_k\to\infty}} |\dot{S}_{k,\xi}(t = t_k)| \le \left(\frac{r}{r+1}\left(1 - \frac{1}{2^{N_s-k-1}}\right) + \frac{1}{r+1}\left(1 - \frac{1}{2^{k-1}}\right) \right) v_{\max}$$

$$\le \left(1 - \min\left\{ \frac{1}{2^{N_s-k-1}}, \frac{1}{2^{k-1}} \right\}\right) v_{\max}$$

$$= v_{\max} - \underbrace{\min\left\{ \frac{1}{2^{N_s-k-1}}, \frac{1}{2^{k-1}} \right\} v_{\max}}_{=\delta>0} \quad (\blacksquare).$$

Theorem A.3 *Suppose that Assumptions 4.1 and 4.2 are satisfied and $v_{max} > 0$. Then, the condition $\max |\dot{q}_{k,\xi}(t)| \le v_{max}$ is satisfied after a finite number of repeat loops by using Algorithm 4.1.*

Proof Theorem A.3 can be proved by using contradiction as in Theorem A.2. Assume that the condition $\max \dot{q}_\xi(t) \le v_{\max}$ is not satisfied as the number of repeat loop tends to infinity. There is a spline segment $S_{k,\xi}$ such that

$$\max |\dot{S}_{k,\xi}(t)| > v_{\max}, \tag{A.28}$$

for arbitrarily large h_k. In Lemma A.1, one has shown that there exist a h_k^* such that

$$\max |\dot{S}_{k,\xi}(t)| = \max\{|\dot{S}_{k,\xi}(t = t_{k-1})|, |\dot{S}_{k,\xi}(t = t_k)|\} > v_{\max}, \quad \forall h_k \ge h_k^*,$$

or

$$\lim_{h_k \to \infty} \max_{t_{k-1} \le t \le t_k} |\dot{S}_{k,\xi}(t)| = 0,$$

which obviously lead to a contradiction of $v_{\max} > 0$. Without the loss of generality, assume that

$$\max\left\{|\dot{S}_{k,\xi}(t = t_{k-1})|, |\dot{S}_{k,\xi}(t = t_k)|\right\} = |\dot{S}_{k,\xi}(t = t_k)| > v_{\max}, \tag{A.29}$$

for arbitrarily large h_k. By applying Algorithm 4.1, if the condition

$$|\dot{S}_{k,\xi}(t = t_k)| \le v_{\max}$$

is not fulfilled, then the interval h_k and h_{k+1} tend to infinity. From Lemma A.2, there exists $\delta > 0$ such that

$$\lim_{\substack{h_{k+1} \to \infty \\ h_k \to \infty}} |\dot{S}_{k,\xi}(t = t_k)| \le v_{\max} - \delta,$$

which is obviously a contradiction to (A.29). Therefore, the condition

$$|\dot{q}_{k,\xi}(t)| \le v_{\max}, \quad \forall k = \{0, 1, \ldots, N_s\}$$

is satisfied after a finite number of loop by using Algorithm 4.1 ($\blacksquare$).

Finally, one can accomplish the proof of Theorem 4.1 presented in Chapter 4 in the following.

Proof (**The termination of Algorithm 4.1**)
Since Theorem A.1, Theorem A.2 and Theorem A.3 are true, it is clear that the Algorithm 4.1 always terminates ($\blacksquare$).

A.2 Heuristic Distance Functions for A* Algorithm

Heuristic Manhattan, infinity norm and Euclidean distance functions between two nodes n_i and n_j are defined as follows:

Manhattan Distance: $h(n_i, n_j) = |x_{n_i} - x_{n_j}| + |y_{n_i} - y_{n_j}|,$

Euclidean Distance: $h(n_i, n_j) = \sqrt{(x_{n_i} - x_{n_j})^2 + (y_{n_i} - y_{n_j})^2},$

Infinity Norm Distance: $h(n_i, n_j) = \max\{|x_{n_i} - x_{n_j}|, |y_{n_i} - y_{n_j}|\},$

where $x_{n_i}, y_{n_i}, x_{n_j}, y_{n_j}$ denote the x and y coordinates of the nodes n_i and n_j.

A.3 Closed Form Solutions of a Cubic Equation

In general, a cubic equation is given by

$$a_3 t^3 + a_2 t^2 + a_1 t + a_0 = 0, \tag{A.30}$$

where $a_3 \neq 0$ and $a_3, a_2, a_1, a_0 \in \mathbb{R}$. In the following, the closed form solutions of the cubic equation are obtained by using the trigonometric method [1]. Since a_3 is non-zero, the equation can be reformulated as

$$t^3 + A t^2 + B t + C = 0, \tag{A.31}$$

where

$$A = \frac{a_2}{a_3}, \quad B = \frac{a_1}{a_3} \quad \text{and} \quad C = \frac{a_0}{a_3}.$$

Let define

$$Q = \frac{-3B + A^2}{9}, \tag{A.32a}$$

$$R = \frac{-9AB + 27C + 2A^3}{54}, \tag{A.32b}$$

$$\Delta = Q^3 - R^2. \tag{A.32c}$$

There are three following cases:

- If $\Delta < 0$: there are one real solution and two conjugate imaginary solutions.

- If $\Delta = 0$: there are three real solutions. At least two solutions are identical.

- If $\Delta > 0$: there are three real and distinct solutions.

If the discriminant Δ is not negative as $\Delta \geq 0$ and let define

$$\cos(\theta) = \frac{R}{\sqrt{Q^3}}, \tag{A.33}$$

then the three real solutions are determined as

$$t_1 = -2\sqrt{Q} \cos\left(\frac{\theta}{3}\right) - \frac{A}{3}, \tag{A.34a}$$

$$t_2 = -2\sqrt{Q} \cos\left(\frac{\theta}{3} + \frac{2\pi}{3}\right) - \frac{A}{3}, \tag{A.34b}$$

$$t_3 = -2\sqrt{Q} \cos\left(\frac{\theta}{3} + \frac{4\pi}{3}\right) - \frac{A}{3}. \tag{A.34c}$$

Otherwise, if the discriminant Δ is negative as $\Delta < 0$ and let define

$$E = \sqrt[3]{\sqrt{-\Delta} - R}, \qquad \text{if } R \leq 0, \tag{A.35a}$$

$$E = \sqrt[3]{-\sqrt{-\Delta} - R}, \qquad \text{if } R > 0, \tag{A.35b}$$

then the only real solution is given by

$$t = E + \frac{Q}{E} - \frac{A}{3}. \tag{A.36}$$